Lecture Notes in Energy 5

For further volumes:
http://www.springer.com/series/8874

Hortensia Amaris • Monica Alonso •
Carlos Alvarez Ortega

Reactive Power Management of Power Networks with Wind Generation

Hortensia Amaris
Monica Alonso
Carlos Alvarez Ortega
University Carlos III of Madrid
Av. Universidad
Madrid, Spain

ISBN 978-1-4471-6006-9 ISBN 978-1-4471-4667-4 (eBook)
DOI 10.1007/978-1-4471-4667-4
Springer London Heidelberg New York Dordrecht

Softcover reprint of the hardcover 1st edition 2013

Printed on acid-free paper

Springer is part of Springer Science+Business Media (www.springer.com)

Preface

This monograph is intended for engineers and scientists interested in wind energy resources and its applications on power systems. The monograph focuses on the reactive power management of power networks with huge amount of wind energy generation. Nowadays, wind energy has widely proved to be one of the most competitive and efficient renewable energy sources and, as a result, its use is indeed continuously increasing. As an example, in June 2010, the total installed wind energy capacity around the world was 175,000 MW. The incorporation of wind energy units into distribution networks not only modifies power flows but also, in some situations, could result in under- or over-voltage on specific points of the network. Furthermore, it would be able to increase the cases of power quality problems and produce any type of alterations regarding voltage stability.

Reactive power compensation systems are presented as a good alternative that aims to alleviate problems related to voltage stability. Reactive power planning in large power systems has become a particularly important task in recent years since it is necessary to develop new techniques to solve any problem that may arise.

The process of high wind energy penetration requires an impact analysis of this new technology applied to power systems. In these terms, some countries have already developed grid codes in order to establish the requirements of wind farms into power networks. Moreover, power network planning with high wind energy penetration requires the definition of several factors, such as the best technology to be used, the optimal number of units to be connected and the optimal size to be chosen.

Currently, a variable-speed wind turbine connected to power systems by means of power electronic converters has the ability to supply reactive power to power systems. This capability allows wind turbines to participate in ancillary services as synchronous generators. However, there are few works focusing on the participation of variable-speed wind turbines in reactive power ancillary services.

A revision of the Grid code requirements in different countries is shown in Chap. 1. Chapter 2 contains a description of the reactive power capability from FACTS devices such as Static Var Compensators (SVC); Static Synchronous compensator (STATCOM) and Dynamic Voltage Restorer (DVR). A brief description of wind generators is given in Chap. 3 where the reactive power capability of fixed-speed wind farms, double-fed induction generators and full-power converter technology is analysed.

The usually applied optimization strategies in power systems for reactive power management analysis are described in Chap. 4. Moreover, metaheuristic techniques have come up to be a good alternative to resolve the problem related to optimal management of reactive power, which involves operation, location and optimal size of these units. Among these techniques, genetic algorithms stand out because of their speed of calculation and simplicity. The proposed optimization strategies will enable not only to determine the optimal location of the reactive power compensation units but also the management of the reactive power injection to fulfill the needs of different load and generation scenarios in power systems.

Chapter 5 provides an overview of the voltage stability problem, which turns out to be even more critical in the power networks which are heavily loaded, are faulted, or have insufficient reactive power supply. As a solution, it is shown that wind energy sources coupled to the network through power converters offer the ability to provide a very fast, dynamic Var injection, and thus their optimal allocation in the power network could alleviate voltage instability or even prevent voltage collapse.

Finally, the benefits associated with the optimal implementation of reactive power management in power systems with high wind energy penetration are analysed in Chap. 6.

In all the above-mentioned chapters, the validity of the designs is corroborated by using simulations in realistic power networks in terms of network security standards.

Contents

Chapter 1
Introduction

Efficient energy management in Europe and the larger application of energy derived from renewable sources, together with energy saving concerns, constitute the basis for the compulsory pack of measures that aim to reduce greenhouse gas emissions and fulfill the Kyoto protocol approved by The United Nations Framework Convention on Climate Change. These measures intend to reduce the emission of greenhouse gases by the year 2012 and in following years [1].

According to the International Agency of Energy, wind energy shall supply 14 % of Europe's power production by the year 2030 and will be set to contribute 60 % of the total increase of power generation from 2006 to 2030. Furthermore, the abovementioned agency states that wind energy will be the most widely developed energy source until 2050, reaching a yearly power value to be installed higher than 70.000 MW. Of this given amount, 30 % will be produced by offshore wind farms. This evolution is founded on various sustainability scenarios presented by different programs such as the BLUE scenario, which aims to reduce greenhouse gas emissions within the timeframe of 2005–2050, relying on wind energy to achieve 26 % of this reduction.

Considering this premise, Directive 2009/28/CE [2], which focused on the promotion of using energy from renewable resources, establishes that, by the year 2020, 20 % of the gross final power consumption for each member state of the European Union shall be produced by renewable sources. To fulfill the European standard, this directive sets up individual objectives for each member state.

1.1 Reactive Power and Voltage Stability

Because of the increase in power demand within electric systems, performing and operating conditions thereof are close to its maximum capacity. These operating conditions have led to most of the problems that have arisen concerning voltage stability within the last several years, and many of these problems resulted in voltage collapse. Voltage stability incidents, such as those taking place in British

H. Amaris et al., *Reactive Power Management of Power Networks with Wind Generation*, Lecture Notes in Energy 5, DOI 10.1007/978-1-4471-4667-4_1,

Columbia (1979), Belgium (1982), Sweden (1983), Brittany and Tokyo (1987), are well documented in the existing literature.

Voltage stability, also named load stability, is closely related to a lack of reactive power in the system. In the past, reactive power correction was usually performed by incorporating var compensators, such as capacitor banks. However, voltage stability could now be improved using equipment based on power electronics known as Flexible Alternative Current Transmission Systems (FACTS) in addition to conventional capacitor banks. Primary characteristics of these devices include their capacities to improve a network's voltage profile and a system's dynamic behavior as well as their ability to enhance power quality. The implementation of FACTS devices is typically justified due to their dynamic contribution of reactive power, voltage control and their quick response.

Reactive power sources stand out as the best devices for improving voltage stability within electric systems. The impact of wind energy on power systems is thus focused on several issues related to the security, stability, power quality and operation of power systems.

- All the utilities must keep the voltage supply stable and reliable within specific limits of frequency and magnitude. The connection of wind farms may result in voltage changes. Consequently, some countries have defined a higher short-circuit level at the connection point, normally between 20 and 25 times the wind farm capacity. There are already some examples of the successful operation of power networks with a lower short circuit level [3].
- Power quality is related to the voltage variation and harmonic distortion in the network. However, the incorporation of wind energy in power networks could affect the power supplied to the customers. To reduce this impact, variable speed wind turbines equipped with power electronics are currently widely used in wind energy conversion. Power electronics improve power quality because they electronics converters can be controlled to reduce the harmonic distortion, Flicker or voltage fluctuation.
- The protection system is also affected by wind farms because the incorporation of wind power injection alters power flow so that conventional protection systems might fail under fault situations.
- In the past, the power network was passively operated and kept stable under most circumstances. However, the power network can no longer be passively operated if an increase in wind energy penetration is considered. Recently, new requirements for wind units have been designed to keep power networks stable under several types of disturbances, such as low voltage ride-through capability.

1.2 Reactive Power Grid Code Requirements

To increase wind energy penetration into the power networks while maintaining the continuity and security of the supply, several countries have developed specific grid codes for wind farms [4]. In general, these grid codes focus on power

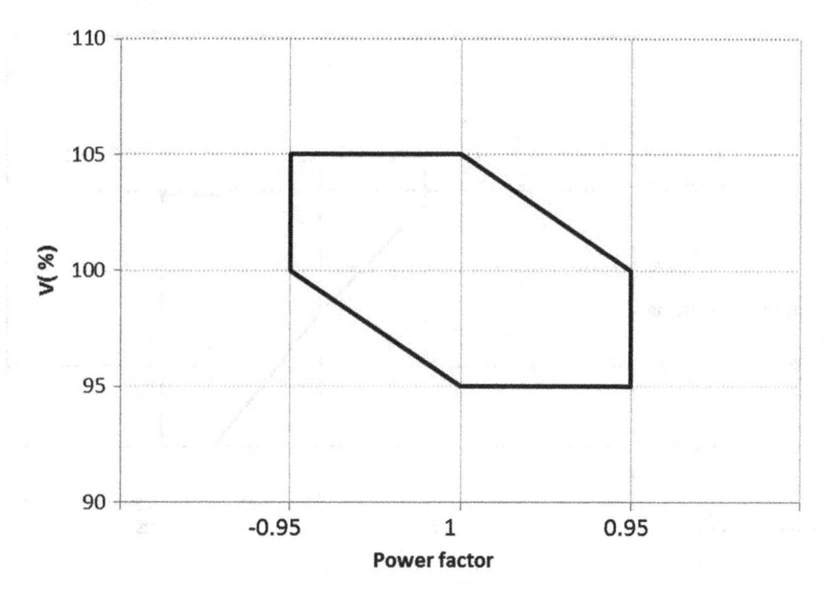

Fig. 1.1 Standard requirements of power factor based on voltage

controllability, power quality and fault ride-through capability where wind turbines are required to offer grid support to the network in case of voltage dips. Figure 1.1 shows the technical requirements of the wind power generators related to the power factor, voltage, and active power.

Under the conditions of voltage variation, grid codes demand that wind generators perform similarly to traditional generators. They must have the capability of providing reactive power whilst providing rated maximum active power.

Figure 1.2 compares those technical requirements based on the power factors of the German grid codes: the E.ON codes [4], [5] and the English grid codes: the NGET codes [6]. The E.ON grid codes are defined for 380, 220 and 110 kV power networks and states that wind farms should be able to work with lagging and leading power factors at the Point of Common Coupling (PCC). The German code has a specific requirement for offshore wind farms [6] where the nominal voltage is 155 kV. In the case of the English codes, NGET requirements [7] establish that wind farms should be able to supply the maximal quantity of reactive power for voltage levels near the nominal value, 1 p.u. at networks with voltage levels 400, 275 and 132 kV. Moreover, the English grid codes require every wind farm to have an automatic voltage control at the connection point. Finally the Irish grid code [8], from ESB National Grid, applies to networks with voltage levels 400, 220 and 110 kV.

Figure 1.3 shows the requirements of the ESB and NGET grid codes related to the variation of reactive power supply capacity. The area delimited by points A, B, C and D corresponds to the Irish requirements regarding reactive power capacity based on the power factor. The black triangle below the 10 % of active power indicates that the reactive power output must be altered during operations below

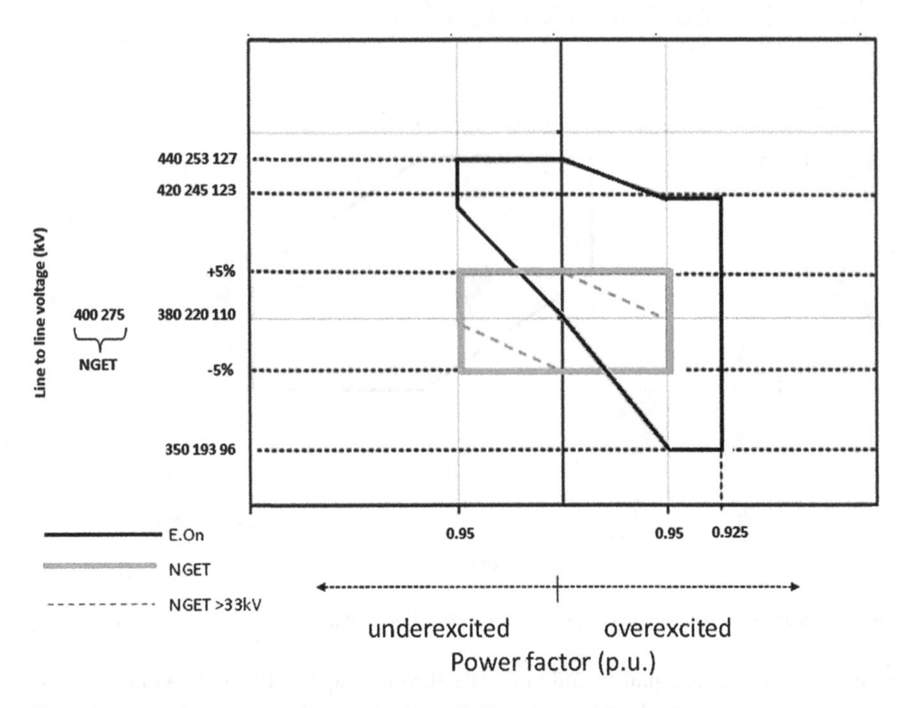

Fig. 1.2 Power factor connection requirements in Germany and England

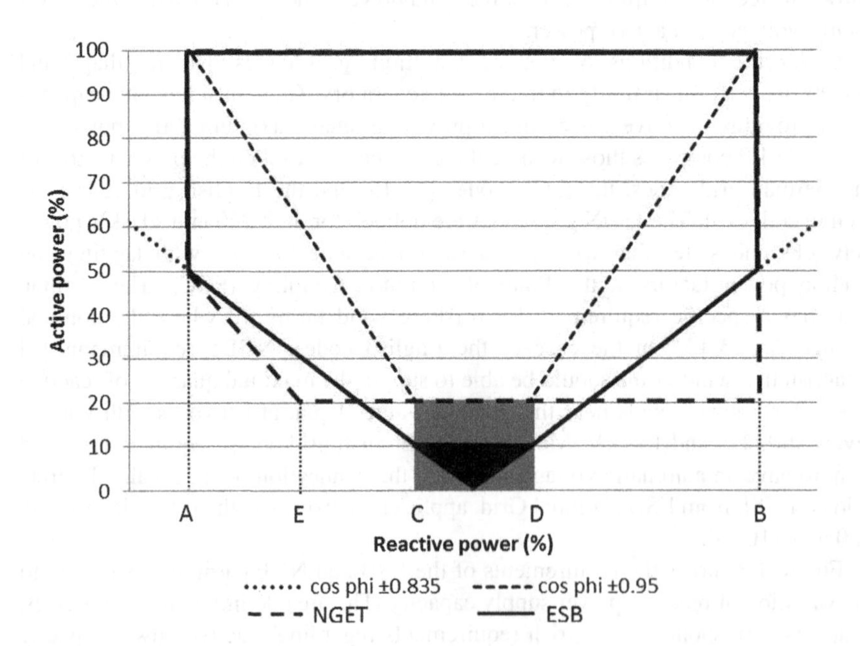

Fig. 1.3 Reactive power requirements in England and Ireland

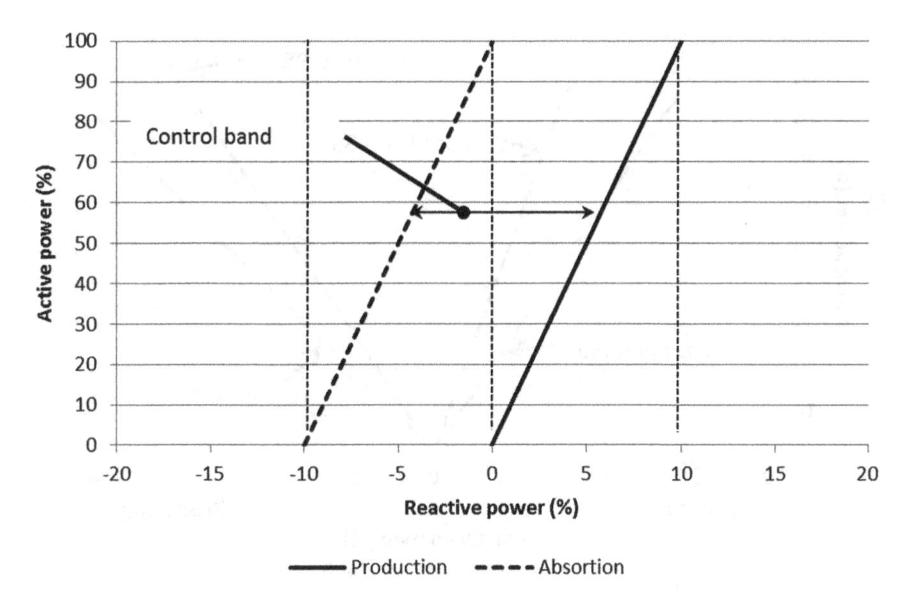

Fig. 1.4 ELTRA reactive power requirements of the wind farm at the connection point

10 %. Point A represents a reactive power supply with a leading power factor of 0.95 at rated active power output while point B corresponds to a performance with a lagging power factor of 0.95 at rated active power output. Similarly, points C and D represent the power factor for a reactive power supply of ±5 %, and point E represents the power factor for 12 % of the nominal capacity. For the performance required by the Irish grid code, generators should supply at least 50 % of their nominal power for a power factor of 0.835, whether a lagging or leading one.

The Nordel grid code [9] states the grid code requirements for the four Scandinavian Countries: Sweden, Denmark, Norway and Finland. Nordel grid code specifies that wind generators should control their reactive power output of the machine to regulate the voltage at the connection point in order to operate continuously on their rated active operation point providing a power factor of:

- 0.95 Whether voltage at the PCC is in the range of 90–100 % rate value.
- 0.95 overexcited if voltage at the PCC is in the range of 100–105 % rate value.

The Eltra grid codes [10] and [11], specify that the reactive power injected by the generators should be able to be modified throughout the range of values shown in Fig. 1.4. The lines in the figure correspond to a power factor of 0.995 where reactive power is considered as a mean value over 10 s. Reactive power control could be performed individually, upon each machine, or upon the whole wind park.

The Belgian transmission system operator ELIA System Operator S.A is in charge of the power networks with voltage levels 380, 220, 150 kV and 94 % of 30–70 kV distribution power networks [12]. Belgian grid codes establish that wind parks with a capacity greater than 25 MW should be able to control reactive power within the interval ranging from −10 % to 45 % of their capacity. In other words,

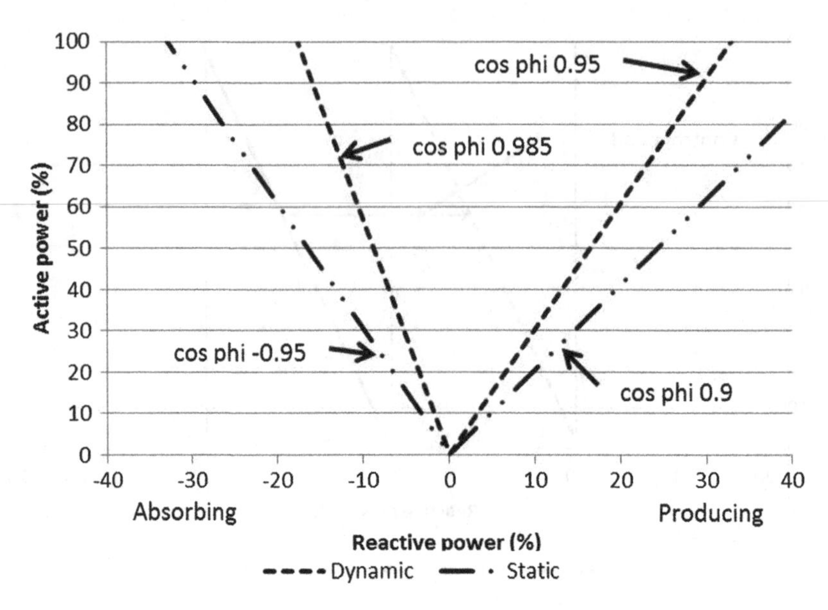

Fig. 1.5 Requirements of the AESO Canadian grid codes

the wind parks must have the capacity to absorb reactive power from or inject reactive power to the system [4].

Hydro Quebec [13] states that wind parks with a capacity greater than 10 MW must be equipped with a voltage automatic control able to operate within the interval corresponding to a rated power factor lagging or leading of 0.95. Moreover, this code reiterates the need for wind parks to contribute to voltage control not only under normal system operation situations but also under dynamic variation.

The Canadian grid codes, the AESO codes [14], states that voltage regulation and reactive power regulation by wind parks must be performed at the low voltage side of the transformers connected to the network. Figure 1.5 shows the requirements concerning reactive power of the AESO Canadian connection grid code. The intent of voltage regulation requirements is to achieve reasonable response to disturbances as well as a steady-state regulation of ± 0.5 % of the controlled voltage [14]. Two areas are to be distinguished: the one corresponding to the system's continuous performance in which the variation range of the power factor is situated within the interval between −0.95 leading and 0.9 lagging at rated active power output and a second area in which reactive power performance happens to be dynamic and the variation interval of the power factor is between −0.985 and 0.95 at rated active power output.

Figure 1.6 shows a brief summary of all the requirements related to the power factor for the grid connection network codes of different analyzed states.

It can be seen that at the rated active power output the Canadian grid code is the more demanding because wind turbines are required to offer a reactive power support from 0.90 lagging to 0.95 leading. This restriction is maintained from

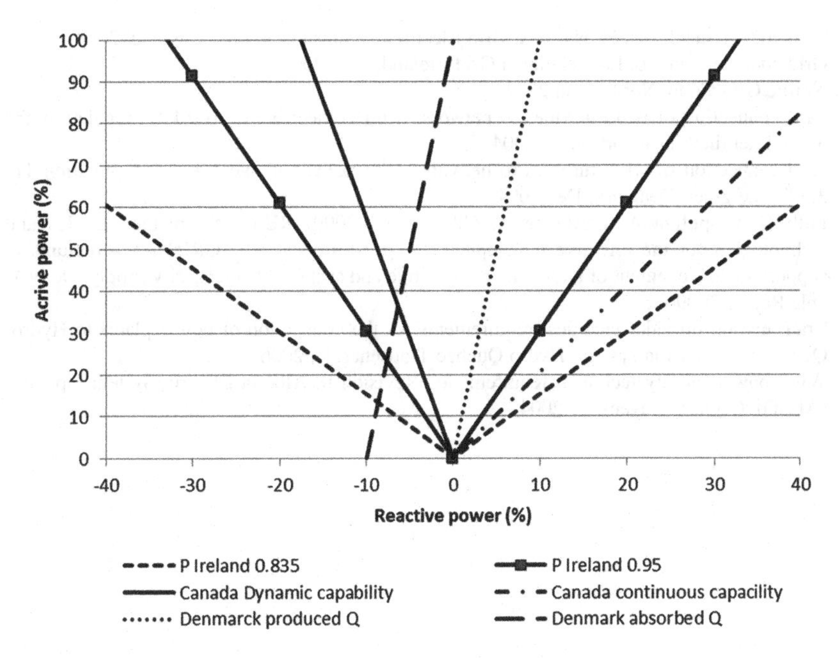

Fig. 1.6 Summary of the requirements depending on different grid codes

100 % to 70 % active power production. Meanwhile, at reduced load situations the Irish grid code will surpass the restrictive Canadian regulations being necessary to offer a reactive power regulation below 0.95 leading and 0.90 lagging.

The fulfillment of the technical requirements regarding the network connection of the fixed-speed wind generators, as far as the reactive power control is concerned, is to be achieved by connecting the FACTS devices at the machine's or the wind park's terminals. In the case of doubly fed asynchronous generators or full power converter, reactive power control could be performed by regulating the electronic converters of the variable speed wind turbine.

References

1. Bianchin R, Azau S (2011) EU energy policy to 2050: achieving 80–95% emissions reductions. EWEA. March 2011
2. DIRECTIVE 2009/28/EC OF THE EUROPEAN PARLIAMENT AND OF THE COUNCIL on the promotion of the use of energy from renewable sources and amending and subsequently repealing Directives 2001/77/EC and 2003/30/EC. 23 Apr 2009
3. Anaya-Lara O, Jenkins N, Ekanayake J, Cartwright P, Hughes M (2011) Wind energy generation: modelling and control. Wiley, Chichester
4. Tsili M, Papathanassiou S (2009) Review of grid code technical requirements for wind farms. IET Renew Power Gener 3(3):1–25
5. EON Grid Code – High and extra high voltage. April 2006
6. Requirements for Offshore Grid Connections in the E.ON Netz Network. April 2008

7. Grid code, issue 3, rev.24. National Grid Electricity Transmission plc, UK, Oct 2008
8. Grid code-version 3.0. ESB National Grid, Ireland, Sept 2007
9. Nordic Grid Code. Nordel, Jan 2007
10. Grid connection of wind turbines to networks with voltage below 100 kV, Regulation TF. 3.2.6. Energinet, Denmark, May 2004
11. Grid connection of wind turbines to networks with voltage above 100 kV, Regulation TF. 3.2.5. Energinet, Denmark, Dec 2004
12. Fulli G, Ciupuliga A, L'Abbate A, Gibescu M (2009) REseArch, methodoLogIes and technologieS for the effective development of pan-European key GRID infrastructures to support the achievement of a reliable, competitive and sustainable electricity supply. Deliverable Report 2009
13. Transmission provider technical requirements for the connection of power plants to Hydro-Quebec transmission system. Hydro Quebec Transenergie, 2006
14. Wind power facility technical requirements. Revision 0. Alberta Electric System Operator (AESO), Canada, November, 2004

Chapter 2
Facts Devices

In recent years technological advances in power electronics have facilitated the development of electronic equipments that offer the ability to handle large amounts of power; consequently, the use and application of this technology into electrical power systems have increased significantly. These electronic devices, called Flexible AC Transmission System (FACTS), are based on electronic power converters and they provide the ability to make quick adjustments and to control the electrical system. FACTS devices can be connected in series, in parallel, or in a combination of both. The benefits they offer to the electrical grid are widely referenced in scientific literature. These benefits include improvement of the stability of the grid, control of the flow of active and reactive power on the grid, loss minimization, and increased grid efficiency.

The installation of FACTS devices (with serial or parallel connections) in a wind farm substation or in the terminals of wind turbines is increasing rapidly owing mainly to the specifications listed in the Transmission System Operators' (TSO) grid codes which require that wind turbines should provide ancillary services similar to those of conventional synchronous generators.

2.1 Static Var Compensator (SVC)

According to the IEEE definition, a Static Var Compensator (SVC) is a shunt-connected static var generator or absorber whose output is adjusted to exchange capacitive or inductive current to maintain or control specific parameters of the electrical power system (typically, the bus voltage) [1].

Typical SVCs can be classified on Thyristor-Controlled Reactor (TCR), Thyristor-Switched Reactor (TSR) or Thyristor-switched capacitors (TSCs). Figure 2.1 shows a TCR single-phase equivalent circuit in which the shunt reactor is dynamically controlled from a minimum value (practically zero) to a maximum value by means of conduction control of the by-directional thyristor valves. By this controlled action the SVC can be seen as a variable shunt reactance established by the parallel connection of

H. Amaris et al., *Reactive Power Management of Power Networks with Wind Generation*, Lecture Notes in Energy 5, DOI 10.1007/978-1-4471-4667-4_2,

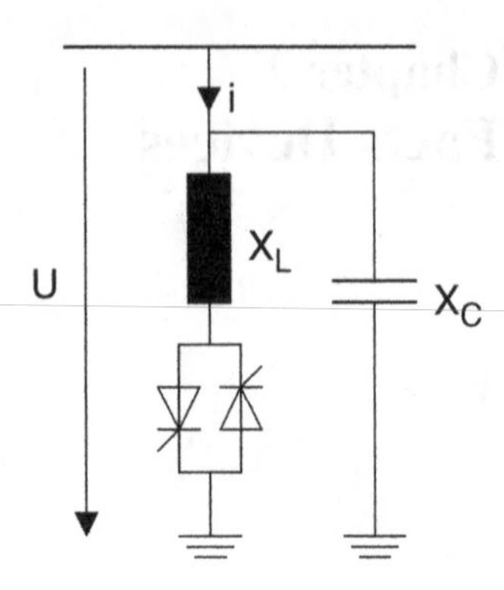

Fig. 2.1 Single-phase equivalent circuit of the shunt SVC (TCR)

the shunt capacitive reactance X_C and the effective inductive reactance X_L controlled by the thyristor switching.

2.1.1 Mode of Operation

The instantaneous current supplied by SVCs is given by:

$$I = \begin{cases} \frac{\mathrm{U}}{\mathrm{X_L}}(\cos\alpha_{svc} - \cos\omega t) & \alpha_{svc} \leq \omega t \leq \alpha_{svc} + \varepsilon \\ 0 & \alpha_{svc} + \varepsilon \leq \omega t \leq \alpha_{svc} + \pi \end{cases} \quad (2.1)$$

Where:

U represents the SVC voltage at the Point of Common Coupling (PCC).
$\mathrm{X_L}$ is the SVC total inductance.
α_{svc} is the firing delay angle.
ε is the SVC conduction angle given by

$$\varepsilon = 2(\pi - \alpha_{svc}) \quad (2.2)$$

It can be seen that as the delay angle α_{svc} increases, the conduction angle $\in$ of the valve decreases.

Figures 2.2, 2.3, and 2.4 show different current wave-shapes injected by the TCR for diverse firing delay angles.

The SVC's control of the output current is based on the control of the firing delay of thyristors. Hence, the maximum injected current is obtained by a firing delay of 90° (full conduction). Meanwhile, the firing angle delays between 90° and 180° electrical degrees only indicate a partial current contribution. This fact contributes to enhancing the device's inductance and makes it possible, at the same time, to decrease its contribution of reactive power and current.

The fundamental component of current is obtained by means of a Fourier analysis (2.3) or in a reduced version (2.4):

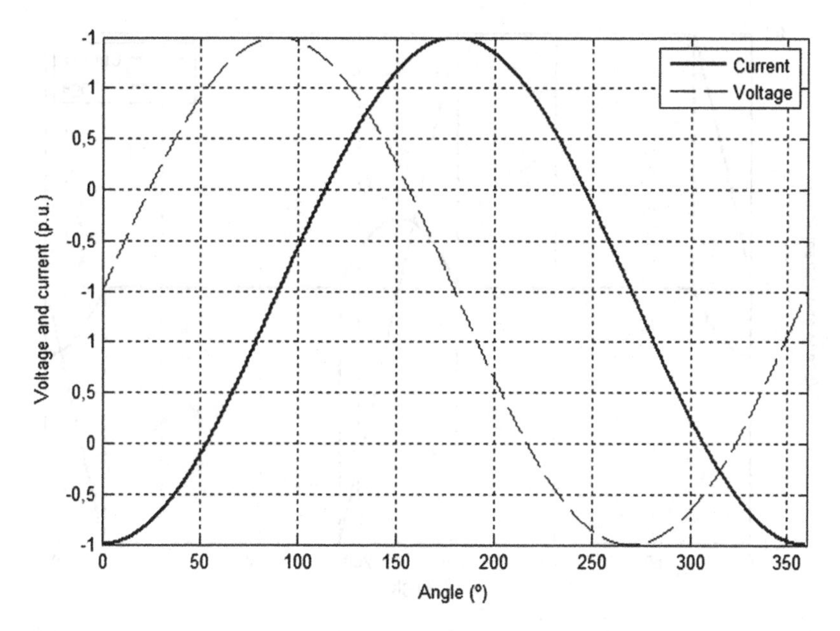

Fig. 2.2 AC current wave shape for $\alpha_{svc} = 90°$

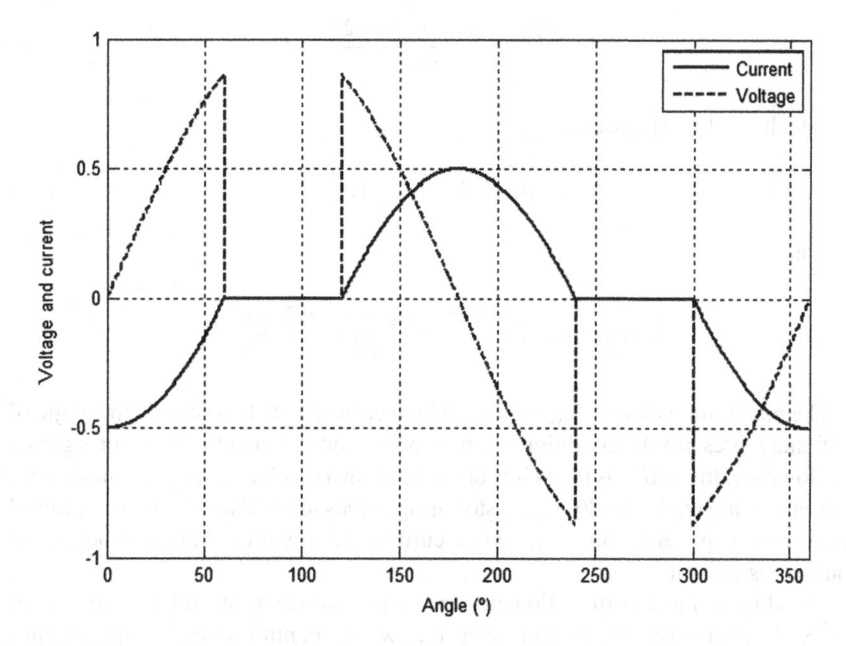

Fig. 2.3 AC current wave shape for $\alpha_{svc} = 120°$

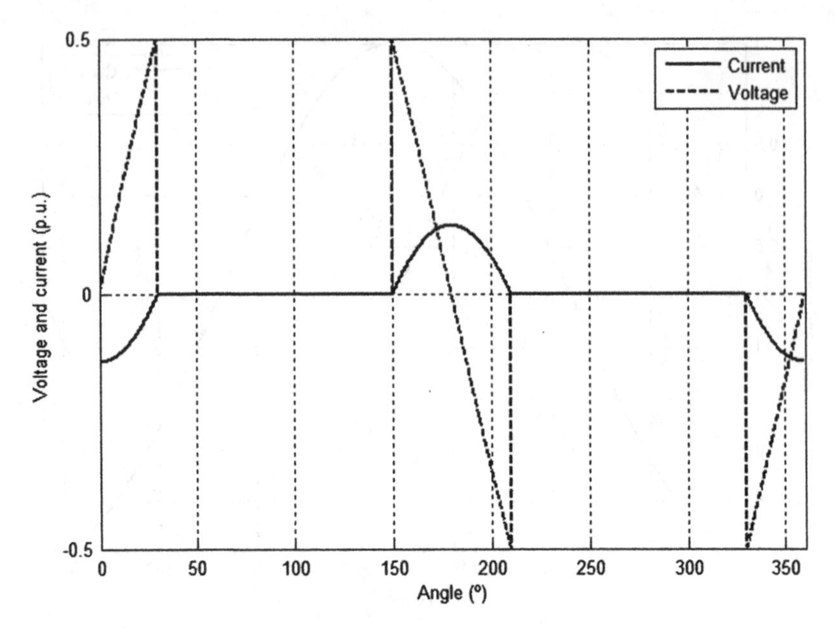

Fig. 2.4 AC current wave shape for $\alpha_{svc} = 150°$

$$I_1 = \frac{2(\pi - \alpha_{SVC}) + \sin 2\alpha_{SVC}}{\pi X_L} U \tag{2.3}$$

Which can be expressed as:

$$I_1 = B_{SVC}(\alpha_{SVC})\mathrm{U} \tag{2.4}$$

Where:

$$B_{SVC}(\alpha_{SVC}) = \frac{2(\pi - \alpha_{SVC}) + \sin 2\alpha_{SVC}}{\pi X_L}$$

The maximum value of B_{svc} is $1/X_L$, which corresponds to a conduction angle of 180° and represents the condition for maximum conduction of the thyristor's group. The lowest value of B_{svc} is 0 and it is obtained from a conduction angle equal to zero or from a firing angle of 180°. Thyristor firing angles lower than 90° are not allowed because they generate an asymmetric current wave with a high component of continuous current.

The slope defined as the ratio between the voltage variation and the variation of the SVC compensating current over the whole control range could be thus expressed via the voltage–current characteristics:

$$U = U_{ref} + X_{SL} I \tag{2.5}$$

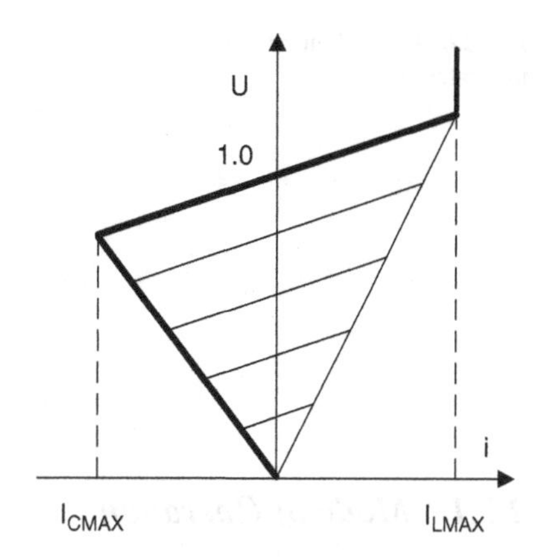

Fig. 2.5 Voltage–current characteristics of the SVC

Typical values of X_{SL} are located within the interval ranging from 0.02 to 0.05 p.u., referring to the SVC's base magnitude. Under limit conditions the SVC will become a fixed reactance. Figure 2.5 shows the characteristic curve of an SVC.

The reactive power supplied by the TCR could thus be calculated using (2.6):

$$Q_{SVC}(\alpha_{SVC}) = \frac{U^2}{X_C} - U^2 B_{SVC}(\alpha_{SVC}) \tag{2.6}$$

From the point of view of power system planning, localization and optimum sizing of FACTS devices are the most important aspects for operating electric networks with high wind power penetration while at the same time maintaining the security and efficiency of the whole electric system.

2.2 Static Synchronous Compensator (STATCOM)

The concept of the STATCOM was proposed by Gyugyi in 1976. According to IEEE a STATCOM can be defined as a static synchronous generator operated as a shunt-connected static var compensator whose capacitive or inductive output current can be controlled independent of the AC system voltage.

A STATCOM is a static compensator that it is connected to the grid in parallel for the compensation of reactive power. It is able to inject or absorb reactive power in a controlled way regardless of the grid voltage [1, 2]. The basic element is the Voltage Source Converter (VSC) which converts an input DC voltage to an AC voltage at the fundamental frequency with a given magnitude and a controllable phase. The AC output voltage is dynamically controlled in order to provide the required reactive power to the network.

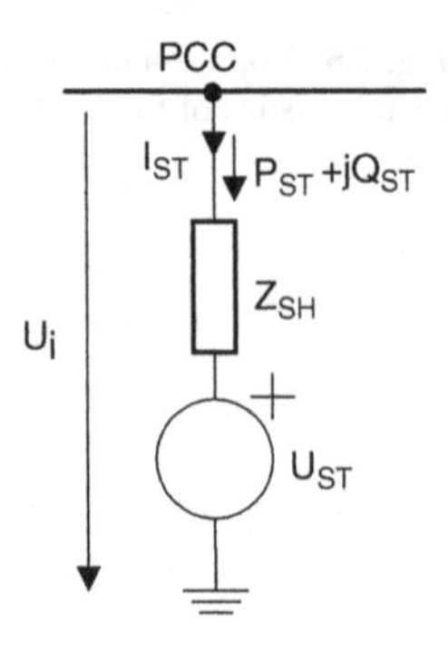

Fig. 2.6 Equivalent circuit of the statcom

2.2.1 Mode of Operation

The VSC generates a voltage at the fundamental frequency $\underset{\rightarrow st}{U} = U_{st}\angle\delta_{st}$ with controllable voltage amplitude and phase. The VSC is connected to the grid $\underset{\rightarrow i}{U} = U_i\angle\delta_i$ through an inductive impedance, Z_{sh}, that represents the coupling transformer and the connection filters, the equivalent circuit is as shown in Fig. 2.6.

The interchange of active and reactive power with the grid can be expressed as follows:

$$P_{st} = U_i^2.g_{sh} - U_iU_{st}[g_{sh}.\cos(\delta_i - \delta_{st}) + b_{sh}.\sin(\delta_i - \delta_{st})] \tag{2.7}$$

$$Q_{st} = -U_i^2.b_{sh} - U_iU_{st}[g_{sh}.\sin(\delta_i - \delta_{st}) - b_{sh}.\cos(\delta_i - \delta_{st})] \tag{2.8}$$

where $Y_{sh} = \frac{1}{Z_{sh}} = g_{sh} + jb_{sh}$.

The capacity for injecting reactive power into the grid is limited by the maximum voltage and the maximum current allowed by the semiconductors, as shown in Fig. 2.7.

The principle of operation of the VSC-based STATCOM depends on the control strategy for regulating the interchange of power between the converter and the grid and it depends also on the output AC voltage of the converter. If the magnitude of the voltage of the converter is equal to the voltage of the grid, $U_{st} = U_i$, the interchange of reactive power between the STATCOM and the grid is equal to zero.

In contrast, if the voltage of the converter is less that the grid voltage at the PCC, $U_{st} < U_i$, the STATCOM absorbs reactive power (draws lagging current). However, if the STATCOM is controlled in such a way that the output voltage of the converter is higher than the PCC voltage, reactive power is injected into the grid [3].

In practice, it is also necessary to control the active power exchange of the STATCOM by regulating the phase angle $\delta_{i\_st} = \delta_i - \delta_{st}$ between the voltage at the VSC ($\underset{\rightarrow st}{U} = U_{st}\angle\delta_{st}$) and the voltage at the PCC ($\underset{\rightarrow i}{U} = U_i\angle\delta_i$) so that the VSC absorbs active power from the grid to maintain a constant voltage for the DC-link.

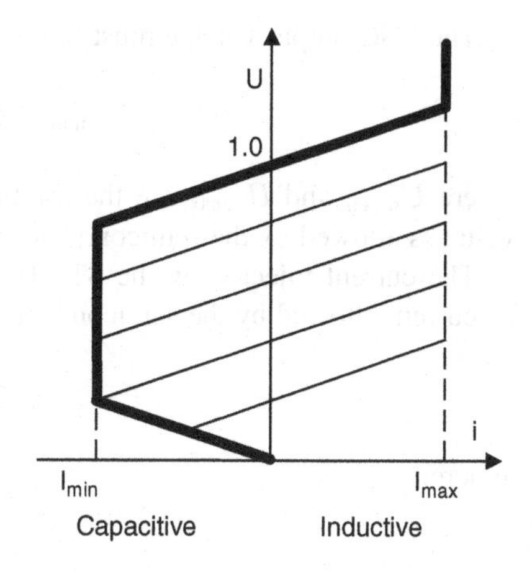

Fig. 2.7 Voltage–current characteristics of the STATCOM

2.2.2 Control Techniques

There are various control techniques, as detailed in [4] where the two listed below are the most typical:

- Voltage local control at the PCC voltage: In this control strategy, the purpose is to regulate the PCC voltage, U_i, so that it is maintained constant at its reference value U_i^{ref}. Mathematically, this condition is expressed as a restriction of operation:

$$U_i - U_i^{ref} = 0 \tag{2.9}$$

- Reactive Power control at the PCC: In many situations, the STATCOM is required to inject reactive power into the grid according to the specifications of the TSO. This form of control can be applied, for example, when a coordinated control is required for FACTS devices and wind farms performing reactive power delivery to the entire grid [5]. When this mode of operation is desired, it must be expressed as a restriction of operation as follows:

$$Q_{st} - Q_{st}^{ref} = 0 \tag{2.10}$$

2.2.3 Restrictions of Operation

In a STATCOM the maximum reactive power that can be supplied to the grid depends on the maximum voltages and currents permitted by the power semiconductors, so it is necessary to include the following internal restrictions:

- The VSC output voltage must fall within the allowed limits of operation:

$$U_{st,\min} \leq U_{st} \leq U_{st,\max} \tag{2.11}$$

where $U_{st,\min}$ and $U_{st,\max}$ are the limiting values of the minimum and maximum voltages allowed by the semiconductors, respectively.

- The current injected by the STATCOM, I_{st}, must be less than the maximum current allowed by the semiconductors, $I_{st,\max}$:

$$I_{st} \leq I_{st,\max} \tag{2.12}$$

Where:

$$I_{st} = \left| \frac{\underset{\rightarrow i}{U} - \underset{\rightarrow st}{U}}{Z_{st}} \right| \tag{2.13}$$

- In contrast, it is necessary to include external restrictions of the grid voltage at the PCC. According to the specific regulations of the grid operator the grid voltage at the PCC must be maintained within certain allowed limits:

$$U_{i,\min} \leq U_i \leq U_{i,\max} \tag{2.14}$$

2.2.4 Application of the STATCOM in Wind Farms

The first studies that analyzed the incorporation of STATCOMs in wind farms were initiated at the end of the 1990s with the aim of improving the flicker and power quality of fixed-speed wind turbines (Fig. 2.8). In these situations, the converter was controlled with a unitary power factor so that there was not any interchange of reactive power between the wind farm and the grid. This strategy has been followed in [6] which demonstrated that by injecting reactive power it was possible to improve the stability and the dynamic operation at the wind farm substation (wind turbines and statcom).

At the Rejsby Hédé wind farm in Denmark a STATCOM was installed in 1998 with a rated power of 8 Mvar for compensating the reactive power of the wind farm. The wind farm consisted of 40 turbines of 600 kW each with a total capacity of 24 MW [7]. The main objective was to improve power quality and to supply the reactive power needed to operate the wind farm. It was noted that it was possible to operate the wind farm with a unitary power factor (instead of regulating the reactive power interchange between the wind farm and the grid) provided that the reactive power demand of the wind farm were inferior to the maximum capacity of the STATCOM; However, above 8 Mvar it was necessary to absorb reactive power from the grid.

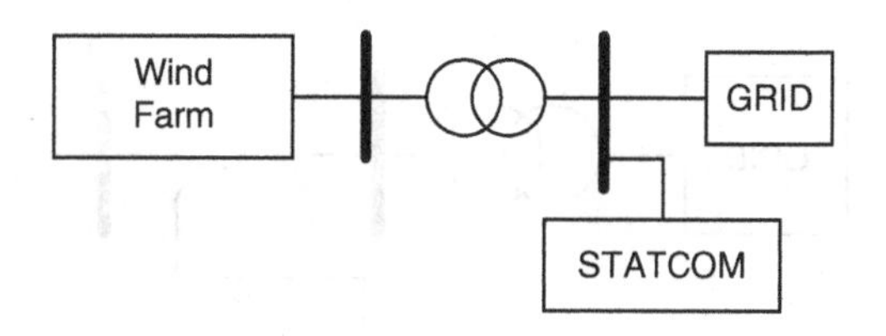

Fig. 2.8 Statcom installed at the wind farm substation

In recent years the new grid code specifications which refer to the regulation of reactive power and Low-Voltage Ride-Through capability (LVRT) have again generated interest in implementing SVCs or STATCOMs in wind farms. In [8] the application of STATCOMs to improve voltage fluctuations is discussed and [9] contains an analysis of how to improve the LVRT capability in fixed-speed wind turbines if STATCOM devices are installed at the wind farm PCC instead of an SVC. This last reference highlights the fact that STATCOMs have an inherent ability to increase the transient stability margin by injecting an adjustable reactive current to the grid regardless of the level of voltage supply; consequently, they are ideal devices for offering LVRT capabilities against voltage dips.

In [10] the possibility of controlling the injection of reactive power of a STATCOM in coordination with DFIG variable-speed wind turbines is analyzed. The main purpose of the STATCOM is to inject reactive power in the PCC in order to reduce the depth of the voltage dip at the wind farm terminal allowing DFIG wind turbines to remain connected during voltage dip situations.

2.3 STATCOM Versus SVC

The main difference between a STATCOM and an SVC is the way they operate: a STATCOM works as a controllable voltage source while an SVC works as a dynamically controllable reactance connected in parallel.

Compared with an SVC, a STATCOM offers the possibility of feeding the grid with the maximum available reactive current even at low voltage levels, this is possible because in every equilibrium condition the injected reactive power varies linearly with the voltage at the Point of Common Coupling (PCC) [11]. In contrast, for an SVC there is a quadratic dependence of the reactive power on the voltage at the PCC which means that to inject the same reactive power it is necessary to install an SVC with a nominal capacity higher than that of a STATCOM.

With regard to the maximum transient capacitive current it is observed that in an SVC the capacitive current is limited by the size of the capacitor and by the magnitude of the AC voltage. In the case of a STATCOM the maximum capacitive current that can be injected is limited by the maximum current capacity of the semiconductors used [12] and is independent of the voltage level at the PCC.

Another feature of a STATCOM is that the DC-link capacitor serves as storage for active power. Therefore in certain situations, depending on the capacitor size, it is possible to regulate the interchange of active power with the grid also.

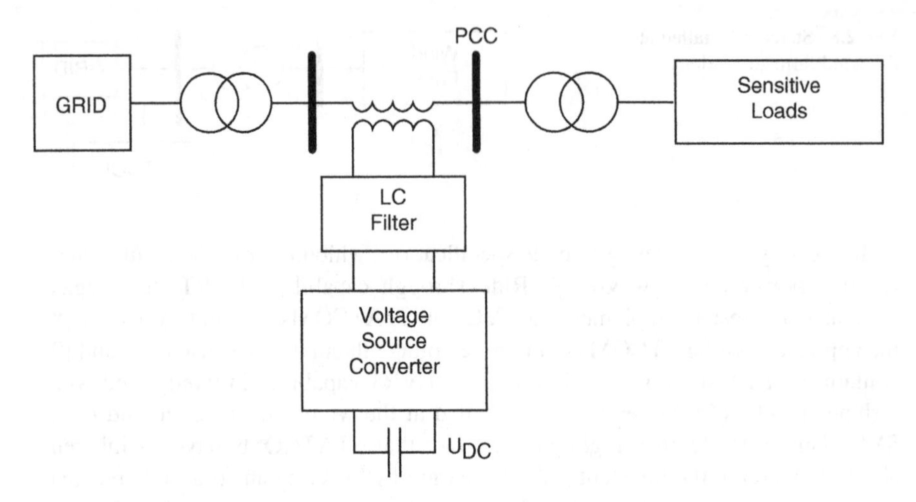

Fig. 2.9 Equivalent circuit of a DVR connected to the grid

STATCOM devices are capable of much faster dynamic reaction (1/4-1 cycle) than an SVC. In a STATCOM the speed of response is limited by the commutation frequency of the IGBT's (normally 1 kHz) [13].

2.4 Dynamic Voltage Restorer (DVR)

A DVR is composed of a Voltage Source Converter (VSC) that has an energy storage connected to the DC link. The VSC is connected in series with the power network by means of a series-connected injection transformer and coupling filters. A DVR may be formed by three VSCs [14] where each one is connected to the network through an LC filter (L_f, C_f) and a transformer. The capacitor filter is connected across the secondary winding of the coupling transformer as shown in Fig. 2.9.

The DVR is normally used to protect critical loads or sensitive installations from the effects of faults at the point of common coupling. During a voltage dip the DVR is able to inject the required voltage to reestablish the load supply voltages.

The typical DVR is based on IGBT solid-state power electronic switching devices in a PWM converter structure and it is capable of independently generating or absorbing controllable real and reactive power at its AC output terminals. For line currents exceeding the inverter rating a bypass scheme can be incorporated to protect the power electronic converter.

DVRs are installed at wind farms mainly for providing low-voltage ride-through capability [14], as is shown in Fig. 2.10.

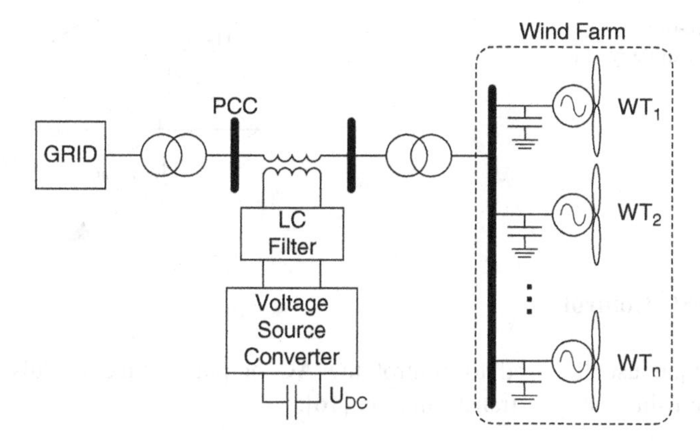

Fig. 2.10 Scheme of a DVR connected at wind farm substation

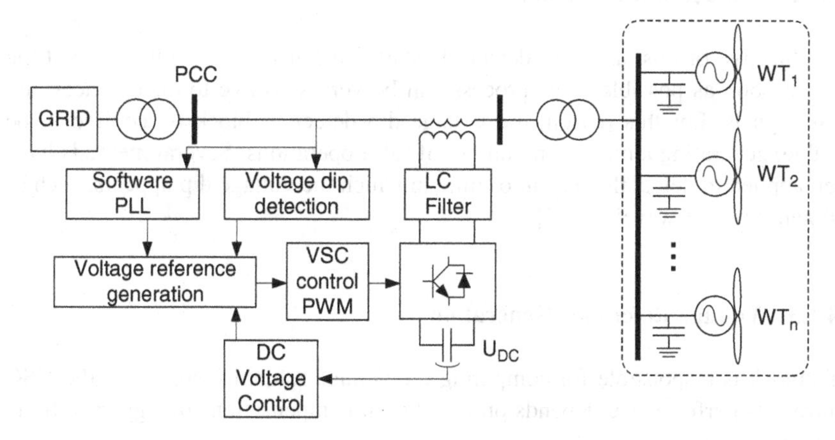

Fig. 2.11 Control blocks at a DVR

2.4.1 DVR Control

The control structure of the DVR includes the following stages (Fig. 2.11): the measured phase voltage before the DVR transformer is fed to the Phase-Locked-Loop (*PLL*) to detect the phase angles and to generate synchronizing signals, U_{ref}. Additionally, the voltage before the DVR transformer is measured to detect and estimate voltage dips. That information is sent to the Voltage reference generation block where it is processed. The result is driven to the VSC Control – PWM, where the switching signals for firing the IGBTs are obtained. The DC voltage is measured for feeding back the Voltage reference generation Control [15].

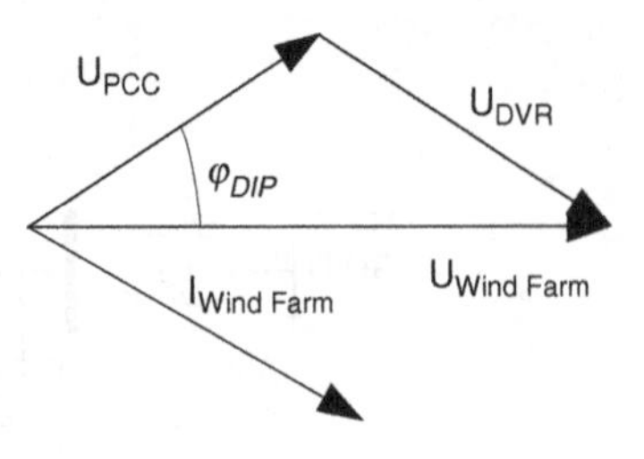

Fig. 2.12 Pre-dip compensation strategy at the DVR

2.4.1.1 VSC Control

The technique usually used to control the AC output voltage is Pulse Width Modulation using the Park transformation [16].

2.4.1.2 Voltage Dip Detection

The objective of this task is to detect the start and finish of the voltage dip at the PCC as soon as possible. This process can be very sensitive to disturbances and noise signals. For this reason, the voltage dip detection block is required to be reliable, generating a minimum number of false operations. Several methods have been applied to detect the instant of time in which the voltage dip appears, such as Kalman filters or wavelets [17].

2.4.1.3 Voltage Reference Generation

This block is responsible for computing the voltage reference signal for the VSC control. Its performance depends on the chosen compensation strategy. The basic strategies are as follows [18]:

- Pre-dip compensation: The simplest solution is to reestablish the exact voltage before the sag (magnitude and phase), see Fig. 2.12. The voltage at the PCC is continuously tracked (U_{PCC}), and with this information, the voltage injected by the DVR (U_{DVR}) is computed to maintain the voltage at the wind farm terminals fixed to the pre-dip situation ($U_{Wind\ Farm}$).

$$U_{DVR} = U_{WindFarm} - U_{PCC} \tag{2.15}$$

- In-phase compensation: This approach relies on compensating the voltage in phase to the grid voltage after the voltage dip. It should be noted that using this compensation strategy makes it possible to minimize the voltage magnitude but the phase jump is not compensated.
- Energy-optimized compensation: This strategy consists of injecting the maximum reactive power by drawing as much active power from the grid as possible. By using this compensation strategy it is possible to restore the voltage but a

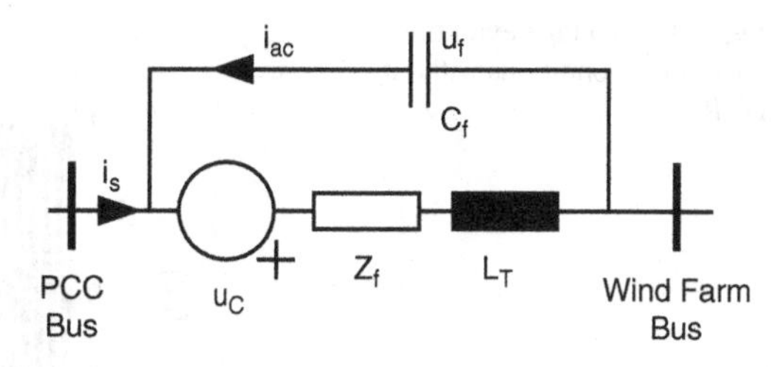

Fig. 2.13 Single-phase equivalent circuit of the DVR

phase jump occurs. This compensation strategy only works properly with small voltage depths.

The single-phase equivalent circuit of the DVR is shown in Fig. 2.13, in which the following parameters are considered:

- u_c corresponds to the switching voltage generated at the AC converter terminals.
- L_T represents the leakage inductance of the series transformer.
- The LC filter is composed of a filter inductance L_f and a capacitor filter C_f. The voltage across the capacitor filter is denoted as u_f.
- The impedance $Z_f = R_{in} + j\omega L_f$ is composed of both the resistance R_{in} which represents the switching losses of the converter and the filter inductance L_f.

From the single-phase equivalent circuit the differential equations of the controller using the state space methodology may be obtained:

$$\dot{x} = \begin{bmatrix} 0 & 1/C_f \\ -1/L_T & -R_{in}/L_T \end{bmatrix} x + \begin{bmatrix} 0 & -1/C_f \\ U_{dc}/L_f & 0 \end{bmatrix} \cdot \begin{bmatrix} u_c \\ i_s \end{bmatrix} \tag{2.16}$$

where the state vector is:

$$x = \begin{bmatrix} u_f \\ i_{ac} \end{bmatrix} \tag{2.17}$$

Equation 2.16 is used to obtain the control law for the series compensator. The controller determines the voltage that has to be injected into the grid to compensate for the voltage dips.

2.4.2 Numerical Results

In this section different simulations are carried out in a wind farm of 20 MVA at a nominal voltage of 11 kV that is composed of fixed-speed wind turbines. Each electrical machine consists of an induction generator of 690 V and 750 kW. Induction generators are provided with bank capacitors for reactive power compensation. The distribution line is represented by its π equivalent circuit and a series

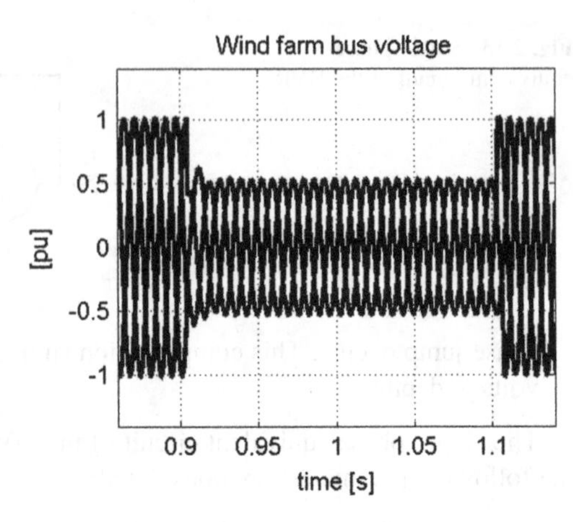

Fig. 2.14 Voltage evolution under fault conditions without DVR

DVR is installed at the wind terminals. Its performance is tested under symmetrical three-faults and unbalanced faults.

The wind farm is assumed to be formed by n generators in parallel all of them with similar characteristics. All the simulations were performed assuming that the whole wind farm is operating at 0.99 leading power factor. For simplicity it has been assumed that each of the turbines experiences the same wind speed, therefore, the whole farm may be represented by its equivalent induction generator.

2.4.2.1 Wind Turbine Performance Under Fault Conditions Without DVR

In this first simulation the DVR is not connected at the wind farm bus. Consequently when a three-phase fault appears at the wind farm terminals (Fig. 2.14) the fixed-speed wind farm is not able to withstand the voltage dip and the angular velocity of the rotor under fault conditions starts to increase, increasing the probability of stability problems, as is depicted in Fig. 2.15.

2.4.2.2 Wind Turbine Performance Under a Three-Phase Fault with DVR

In this scenario the DVR is connected at the wind farm bus. A symmetrical three-phase fault is simulated at the wind terminals with a voltage depth of 50 % and a phase jump of 20° at the beginning of the voltage dip. The voltage dip at the PCC lasts from instant of time t = 0.9 ms to t = 1.1 ms.

In Fig. 2.16 it can be noted that the DVR detects the voltage dip efficiently and injects the required voltage to restore the voltage seen by the wind farm.

It should be highlighted that by using the pre-dip compensation strategy the wind farm's active and reactive power injected to the network is not affected by the voltage dip (Fig. 2.17) and remains almost fixed at the reference values. The angular velocity and the electromagnetic torque of the equivalent wind turbine (Fig. 2.18) recover nominal values after the voltage dip.

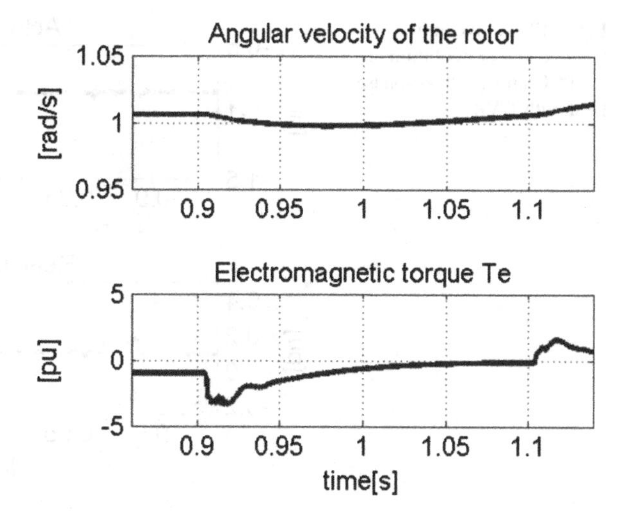

Fig. 2.15 Angular velocity and electromagnetic torque evolution under fault conditions without DVR

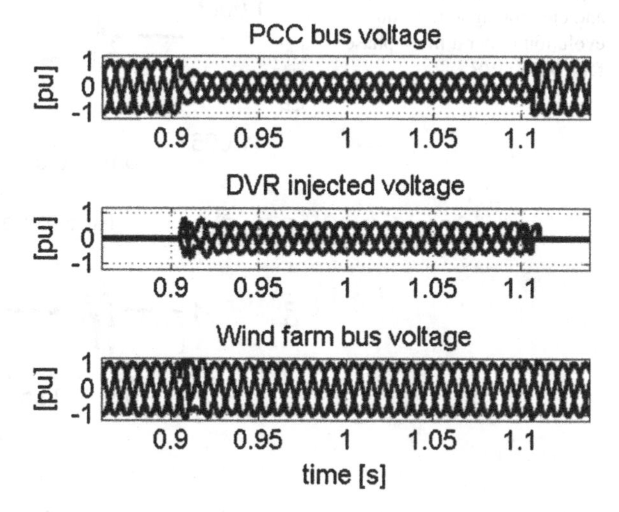

Fig. 2.16 Voltage evolution under a three-phase fault with DVR

2.4.2.3 Wind Turbine Performance Under a Single-Phase to Ground Fault with DVR

In this last case the performance of the DVR has been tested under unbalanced fault conditions in which a single-phase to ground fault has been simulated at the wind terminals. As seen at the PCC voltage (Fig. 2.19) the DVR detects the beginning of the voltage dip, the voltage depth and the phase that is affected. Consequently, it only injects the required compensation voltage in the affected phases.

Both the wind farm injected power and the angular velocity of the equivalent turbine are shown in Figs. 2.20 and 2.21, respectively. Notably, the stability of the wind farm is not affected by the voltage dip.

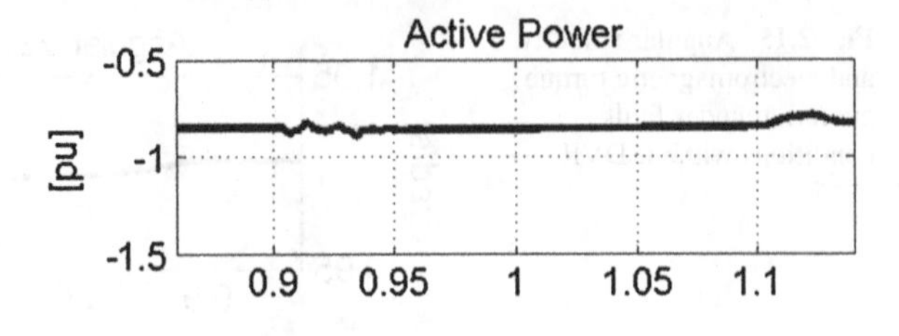

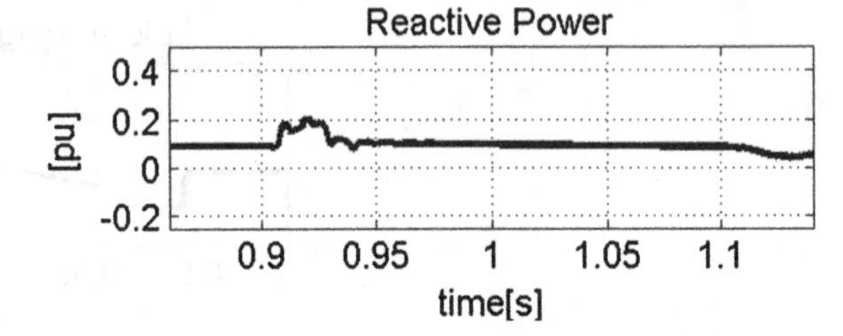

Fig. 2.17 Active and reactive power evolution at the PCC under a three-phase fault with DVR

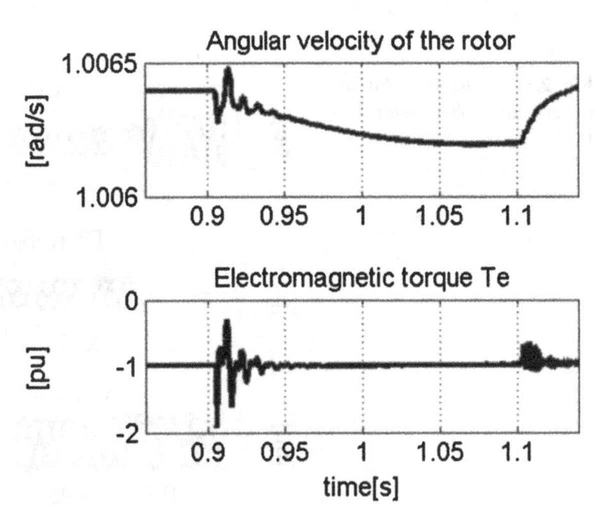

Fig. 2.18 Angular velocity and electromagnetic torque evolution under a three-phase fault with DVR

These two examples show how the installation of a serial DVR at wind farms can efficiently improve the LVRT capability of fixed-speed wind farms under fault situations.

2.4.3 *Reactive Power Support Under Voltage Dips in Fixed-Speed Wind Farms*

Most utilities establish codes for their national grid that state that wind farms shall offer reactive power support to the grid during faults. Consequently during the

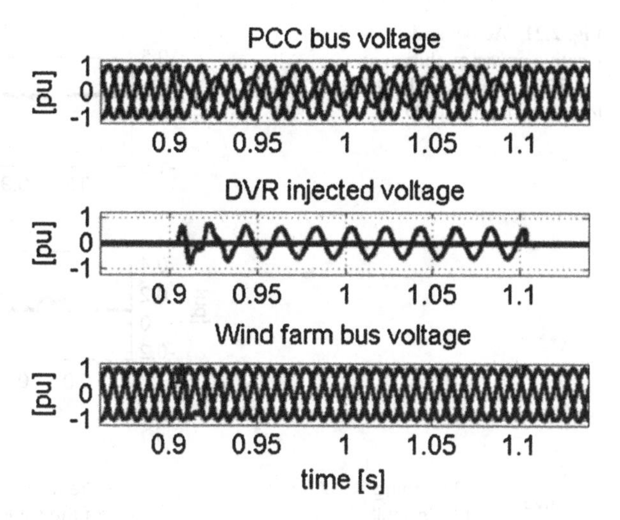

Fig. 2.19 Voltage evolution under a single-phase to ground fault with DVR

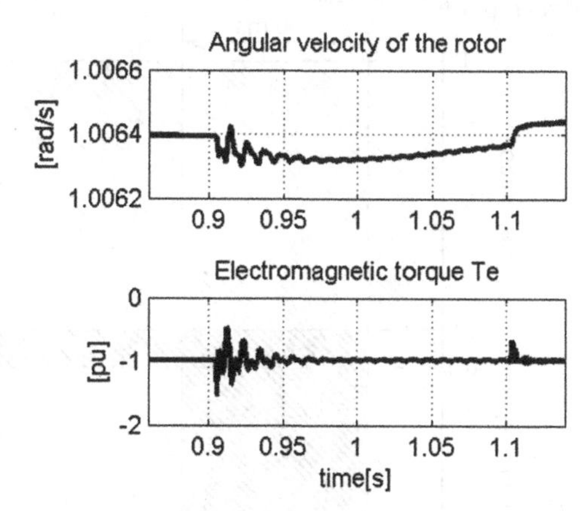

Fig. 2.20 Angular velocity and electromagnetic torque evolution under a single-phase to ground fault with DVR

occurrence of a voltage dip they are required to stay connected and to inject reactive power during both the fault and post-fault periods.

The Spanish grid code for wind turbines [19] establishes the active and reactive power requirements with which the wind turbine must comply during voltage dips (see Figs. 2.22 and 2.23).

Because fixed-speed wind farms do not offer LVRT capabilities by themselves a serial DVR is proposed to offer voltage support at the PCC and to inject the required reactive power during voltage dips.

The operation conditions of the DVR depend on the working conditions of the wind farm. The wind farm working conditions are established by the apparent power $S_{WindFarm}$ generated by the wind farm, as expressed in (2.18). This power will develop a certain current flowing from the wind farm to the DVR series transformer.

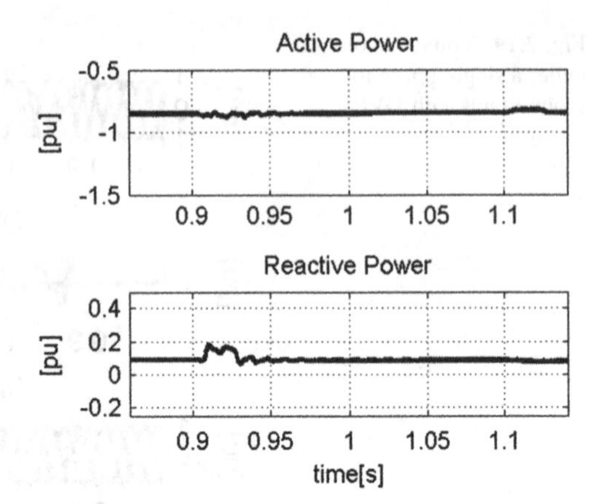

Fig. 2.21 Active and reactive power evolution at the PCC under a single-phase to ground fault with DVR

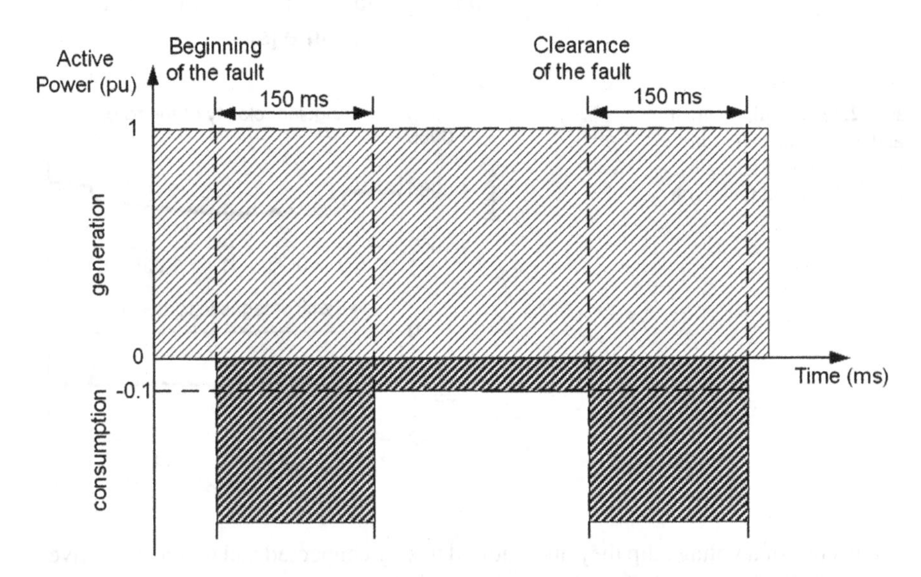

Fig. 2.22 Active power requirements at wind farm substation during voltage dips [19]

$$\underline{S}_{WindFarm} = S_{WindFarm} \angle \varphi_{WindFarm} = P + jQ \tag{2.18}$$

The results of the simulation are shown in Figs. 2.24 and 2.25. In Fig. 2.24 the voltage profile at the PCC, the voltage at the wind farm terminals and the voltage injected by the DVR is depicted, in this figure it can be noted how the voltage at the wind farm holds in a range of voltages around the reference value (1.0 p.u). Figure 2.25 shows how the active power injected by the fixed speed wind farm oscillates during the voltage dip in order to comply with the grid code requirements. The grid code establishes that the active and reactive power at the PCC could be

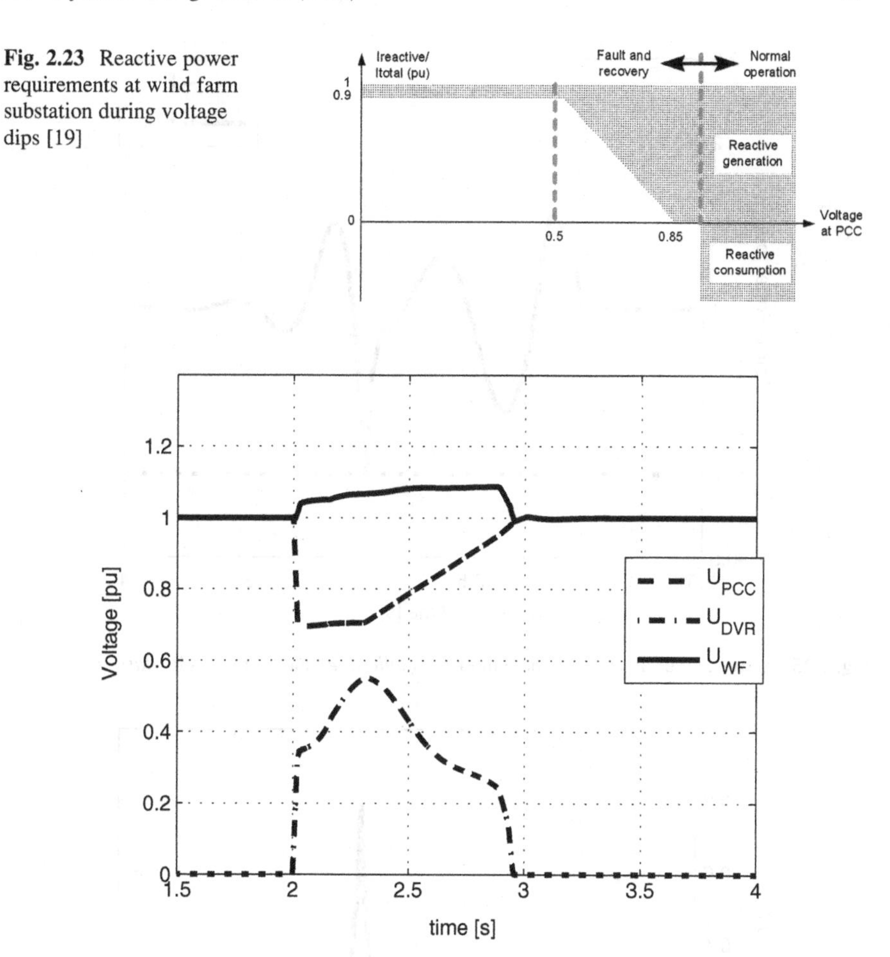

Fig. 2.23 Reactive power requirements at wind farm substation during voltage dips [19]

Fig. 2.24 Voltage at the PCC (*dashed line*), voltage injected by the DVR (*dash-dot line*) and voltage at the wind farm substation (*solid line*)

consumed or generated for a short period of time that was determined to be 150 ms [19]. For the rest of the fault both active and reactive power will be generated.

It must be noted that the wind farm does not provide reactive power to the grid, Fig. 2.26 shows the reactive current injected to the grid by the DVR in order to comply with the reactive power requirements (Fig. 2.23).

Previous results plotted in Figs. 2.24, 2.25, and 2.26 show the good performance of the DVR according to the LVRT requirements in the grid code [19].

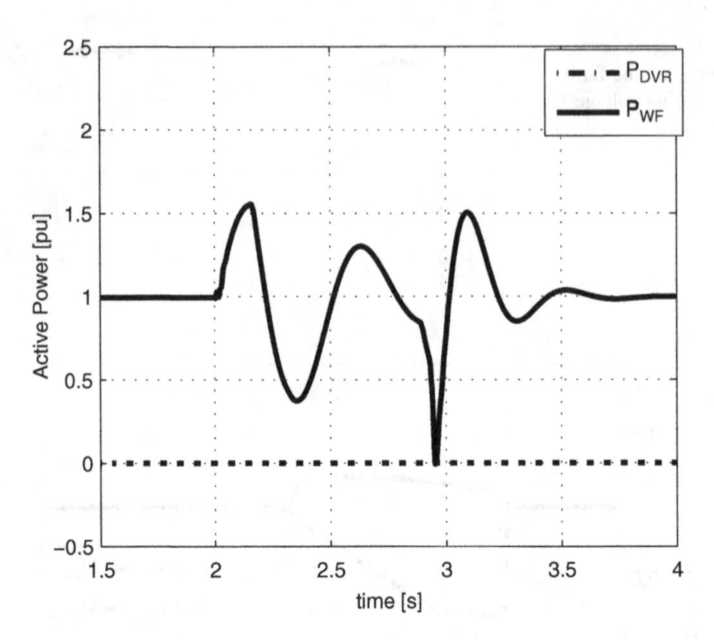

Fig. 2.25 Active power injected by the wind farm (*solid line*) and by the DVR (*dash line*)

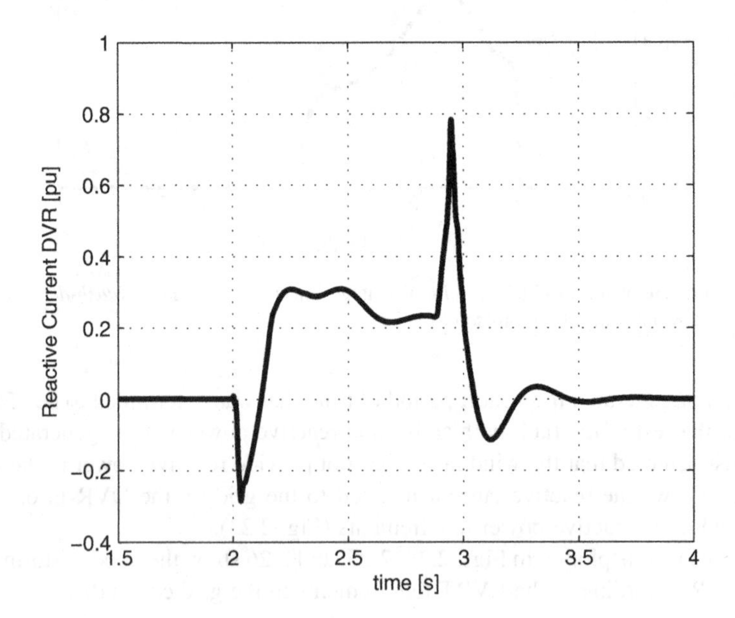

Fig. 2.26 Reactive current injected by the DVR

References

1. Hingorani NG, Gyugyi L (1999) Undestanding FACTS. IEEE Press, New York
2. Mithulananthan N, Canizares CA, Reeve J, Rogers GJ (2003) Comparison of PSS, SVC, and STATCOM controllers for damping power system oscillations. IEEE Trans Power Syst 18 (2):786–792
3. Moore P, Ashmole P (1998) Flexible AC transmission systems. 4. Advanced FACTS controllers. Power Eng J 12(2):95–100
4. Zhang X-P, Rehtanz C, Pal B (2006) Flexible AC transmission systems: modelling and control. Springer, Berlin
5. Amaris H, Alonso M (2011) Coordinated reactive power management in power networks with wind turbines and FACTS devices. Energ Convers Manage 52(7):2575–2586
6. Saad-Saoud Z, Lisboa ML, Ekanayake JB, Jenkins N, Strbac G (1998) Application of STATCOM's to wind farms. IEE Proc Gener Trans Distrib 145(5):511–516
7. Sobtink KH, Renz KW, Tyll H (1998) Operational experience and field tests of the SVG at Rejsby Hede. In: International conference on power system technology proceedings, POWERCON'98, vol 1, Zhejiang, pp 318–322
8. Han C, Huang AQ, Baran ME, Bhattacharya S, Litzenberger W, Anderson L, Johnson AL, Edris A (2008) STATCOM impact study on the integration of a large wind farm into a weak loop power system. IEEE Trans Energ Convers 23(1):1266–1272
9. Molinas M, Are Suul J, Undeland T (2008) Low voltage ride through of wind farms with cage generators: STATCOM versus SVC. IEEE Trans Power Electron 23(3):1104–1117
10. Qiao W, Harley RG, Venayagamoorthy GK (2009) Coordinated reactive power control of a large wind farm and a STATCOM using heuristic dynamic programming. IEEE Trans Energ Convers 24(2):493–503
11. Molinas M, Vazquez S, Takaku T, Carrasco JM, Shimada R, Undeland T (2005) Improvement of transient stability margin in power systems with integrated wind generation using a STATCOM: an experimental verification. In: International conference on future power systems, vol 1, pp 1–6
12. Zhang W (2007) Optimal sizing and location of static and dynamic reactive power compensation. PhD thesis, University of Tennessee, Knoxville
13. Han C, Yang Z, Chen B, Song W, Huang AQ, Edris A, Ingram M, Atcitty S (2005) System integration and demonstration of a 4.5 MVA STATCOM based on emitter turn-off (ETO) thyristor and cascade multilevel converter. In: 31st annual conference of IEEE Industrial Electronics Society, IECON 2005, Raleigh, pp 1329–1334
14. Lam C-S, Wong M-C, Han Y-D (2004) Stability study on dynamic voltage restorer (DVR). In: Proceedings of the first international conference on power electronics systems and applications, Hong Kong, China, pp 66–71
15. Alvarez C, Amaris H (2006) Voltage dips compensation in wind farms using dynamic voltage restorer. In: Proceedings of the 5th international Nordic workshop on power and industrial electronics, NORPIE 2006, Aalborg
16. Nielsen JG, Newman M, Nielsen H, Blaabjerg F (2004) Control and testing of a dynamic voltage restorer (DVR) at medium voltage level. IEEE Trans Power Electron 19(3):806–813
17. Lobos T, Rezmer J, Janik P, Amaris H, Alonso M, Alvarez C (2009) Application of wavelets and prony method for disturbance detection in fixed speed wind farms. Int J Electron Power Energ Syst 31:429–436
18. Meyer C, De Doncker RW, Wei Li Y, Blaabjerg F (2008) Optimized control strategy for a medium-voltage DVR – theoretical investigations and experimental results. IEEE Trans Power Electron 23(6):2746–2754
19. P.O. 12.3. Requisitos de Respuesta frente a Huecos de Tensión de las Instalaciones Eólicas, Resolución del Ministerio de Industria, Oct 2006. (In Spanish)

References

1. Hingorani NG, Gyugyi L [illegible] Understanding [illegible]TS. IEEE Press, [illegible] York
2. Ma[illegible] [illegible] Rogers G [illegible] (2003) [illegible] STATCOM [illegible] Power Syst [illegible] (2):786–792
3. Moore P, [illegible] (1998) [illegible] Power [illegible]
4. [illegible] (2011) [illegible] Flexible AC transmission systems: modelling and control. Springer, Berlin
5. [illegible] (2011) [illegible] power [illegible] in power networks [illegible] and FACTS devices. [illegible] 2(2):[illegible]
6. [illegible] Application of STATCOM [illegible] Power Deliv [illegible]
7. [illegible] (20[illegible]) Operational experience and field tests of the SVC [illegible] power system [illegible] IEEE [illegible]
8. [illegible]
9. [illegible] voltage [illegible] wind [illegible] Power [illegible]
10. [illegible] power control of a [illegible] STATCOM [illegible] (2): [illegible]
11. Molinas M, [illegible] Takashi [illegible] Undeland T (2008) [illegible] of transient [illegible] margin in power system with [illegible] wind generators [illegible] STATCOM: experimental verification [illegible] Power Syst [illegible]
12. Zhang W (2007) [illegible] dissertation, [illegible] Knoxville
13. [illegible] B, [illegible] M, [illegible] (2008) [illegible] STATCOM [illegible] cascade multilevel converter [illegible] Industrial Electronics Society (IECON 2008), Orlando, pp [illegible]
14. [illegible] Proceedings [illegible] electric utility [illegible] power electronics systems [illegible] Hong Kong, China [illegible]
15. [illegible] (2008) [illegible] voltage [illegible] Proceedings of [illegible] International Conference on [illegible] Renewable Energies (ICREPQ 2008), [illegible]
16. [illegible] (2012) [illegible] voltage [illegible] IEEE Trans [illegible] 27(3):[illegible]–813
17. [illegible] Application of [illegible] and [illegible] fixed speed wind farms [illegible] Power Energy [illegible]
18. [illegible] W, [illegible] Y, Bl[illegible] (2011) Optimized control strategy for [illegible] theoretical investigation and experimental results [illegible] Electr Power [illegible]
19. P.O. [illegible] Respuesta frente a Huecos de Tensión de las Instalaciones Eólicas [illegible] Ministerio de Industria [illegible]

Chapter 3
Wind Generators

The behavior of power systems with large wind energy farms requires a detailed examination of the available wind turbine technologies to evaluate the power capabilities of each technology and their operating regions. Wind turbines can be classified between fixed-speed (small speed changes due to generator slip) and variable-speed.

Traditionally, wind farms have been represented as PV or PQ models in power flow studies [1, 2]. The methodology implemented till now seems to be quite simple, however it presents the main drawback that the available reactive power range is limited either to a maximum power factor or to a fixed regulation band [3–5]. Additionally, this representation is not completely accurate and therefore does not allow taking full advantage of the reactive power injection from the wind turbine.

Some grid codes establish a minimum level of reactive power control; this requirement is related to the ability of wind units to work within a power factor range between 0.95 leading and 0.95 lagging. The reactive power capability from variable-speed turbines may differ depending on the variable-speed technology (double-fed induction generator or direct driven generators). Modern wind units use variable-speed generators connected to the grid by power electronic converters. These converters offer the possibility to control the reactive power output of wind units by varying voltage magnitude and frequency. Double Fed Induction Generators (DFIG) are the most popular generators employed in wind units and could offer dynamic reactive power control due to the grid side converters. This chapter analyzes the reactive power capabilities of fixed-speed and variable-speed wind turbines.

H. Amaris et al., *Reactive Power Management of Power Networks with Wind Generation*, Lecture Notes in Energy 5, DOI 10.1007/978-1-4471-4667-4_3,

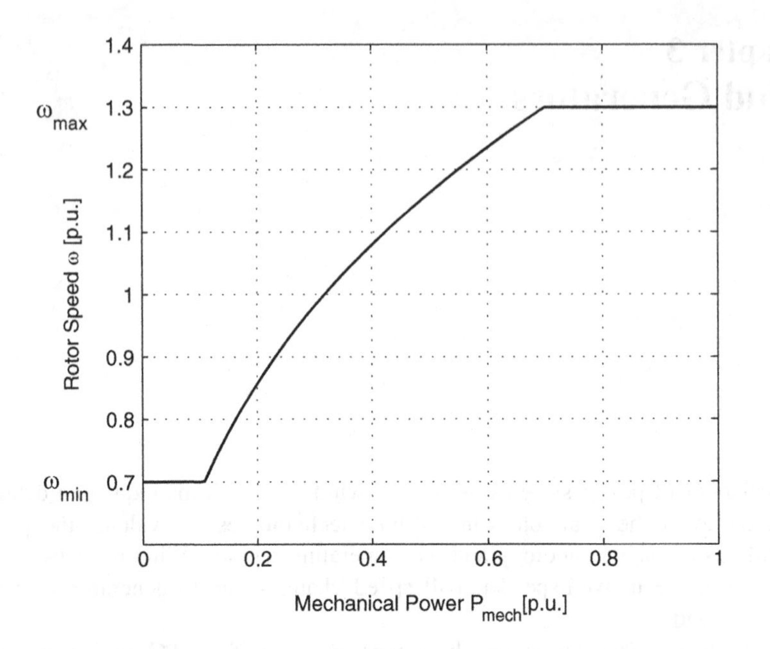

Fig. 3.1 Rotor speed evolution

3.1 Wind Turbine

The available wind power increases with the wind speed [6] according to:

$$P_{mech} = \frac{1}{2}\rho C_p(\lambda_v, \beta)Av^3 \tag{3.1}$$

Where ρ is air density, $\lambda_v = \lambda_v(\omega, v)$ is tip speed ratio, ω is rotational rotor speed, v is wind speed, A is rotor swept area, and C_p is a power coefficient value depending on λ_v and pitch blade angle β.

In the existing literature there are several optimization strategies that could be used to find the working point as a function of wind speed. In this way, (3.1) can be considered as a dependent function on the rotor speed:

$$P_{mech} = P_{mech}(\omega) \tag{3.2}$$

and, consequently, its inverse:

$$\omega = \omega(P_{mech}) \tag{3.3}$$

The generic relationship of (3.3) could be refined in order to include any further restriction, such as: minimum and maximum mechanical speed, security limits as well as other design aspects. A convenient way to use (3.3) is by expressing it as an exponential relation, as shown in Fig. 3.1:

$$\omega = K_S P_{mech}^{\gamma}, \quad \omega_{min} \leq \omega \leq \omega_{max} \tag{3.4}$$

where K_S and γ are constants depending on wind turbine machine design parameters, and ω_{min}, ω_{max} define the limit values for ω variation.

Considering ω_1 as the grid angular frequency, the slip s shall be defined as:

$$s = \frac{\omega_1 - \omega}{\omega_1} \tag{3.5}$$

Therefore, the rotor slip, s, shall be regarded as a dependent function on P_{mech}:

$$s = s(P_{mech}) \tag{3.6}$$

3.2 Mechanical Model

The shaft system represents the coupling between the mechanical and electrical parameters of a wind turbine and can be correctly represented using a two-mass approach. In this two-mass model, the hub and blades are grouped together in the low-speed shaft mass, and the rotor and generator are grouped together in the high-speed shaft, as follows:

$$\frac{d\theta_{tg}}{dt} = \omega_g - \omega_t \tag{3.7}$$

considering:

$$\begin{aligned} \frac{d\omega_g}{dt} &= \frac{1}{2H_t}\left[T_{mec} + k\theta_{tg} + D(\omega_g - \omega_t)\right] \\ \frac{d\omega_t}{dt} &= \frac{1}{2H_g}\left[-T_e - k\theta_{tg} - D(\omega_g - \omega_t)\right] \end{aligned} \tag{3.8}$$

Where:

ω_t, ω_g are the turbine and generator speeds, respectively,
H_t, H_g are the turbine and generator inertia constants, respectively,
θ_{tg} is the angle between the turbine and the generator,
K is the drive train stiffness,
D is the drive train damping,
T_{mec} is the aerodynamic torque provided by the wind,
T_e is the electromagnetic torque from the induction machine.

3.3 Fixed Speed Wind Turbines

Fixed speed wind turbines became the dominant technology during the 1990s [7], they are composed of squirrel induction generators coupled directly to the grid. Consequently, the only ways to control these generators are to adjust the pitch angle of the turbine blades and thereby their aerodynamic efficiency (Cp) or to vary the voltage applied to the stator. In this type of turbine the speed of the generator slips over a limited range of wind speed (usually up to 1 %).

The dynamic behavior of the induction machine can be represented using the detailed fifth-order model [8]. Adopting generator convention and considering stator currents to be positive when flowing towards the grid, the equations are transformed into a direct d-axis and a quadrature q-axis reference frame with the axes rotating at a synchronous speed ω_s.

Stator voltages U_{ds} and U_{qs} are calculated as follows:

$$\begin{aligned} U_{ds} &= R_s i_{ds} - \omega_s \lambda_{qs} + \frac{d\lambda_{ds}}{dt} \\ U_{qs} &= R_s i_{qs} + \omega_s \lambda_{ds} + \frac{d\lambda_{qs}}{dt} \end{aligned} \tag{3.9}$$

and rotor voltages U_{dr} and U_{qr} are calculated as follows:

$$\begin{aligned} U_{dr} &= R_r i_{dr} - s\omega_s \lambda_{qr} + \frac{d\lambda_{dr}}{dt} = 0 \\ U_{qr} &= R_r i_{qr} + s\omega_s \lambda_{dr} + \frac{d\lambda_{qr}}{dt} = 0 \end{aligned} \tag{3.10}$$

where the subscripts (s, r) represent the stator and rotor, respectively; subscripts (d, q) correspond to the components aligned to the d-axis and q-axis rotating frames, respectively; and s is the rotor slip.

Stator flux linkages λ_{ds} and λ_{qs} are calculated as follows:

$$\begin{aligned} \lambda_{ds} &= L_s i_{ds} + L_m i_{dr} \\ \lambda_{qs} &= L_s i_{qs} + L_m i_{qr} \end{aligned} \tag{3.11}$$

and rotor fluxes λ_{dr} and λ_{qr} are calculated as follows:

$$\begin{aligned} \lambda_{dr} &= L_r i_{dr} + L_m i_{ds} \\ \lambda_{qr} &= L_r i_{qr} + L_m i_{qs} \end{aligned} \tag{3.12}$$

The electric parameters of the induction machine R_s, L_s, R_r, L_r, L_m represent the stator resistance and inductance, the rotor resistance and inductance, and the mutual inductance, respectively.

This detailed model includes electromagnetic transients in both the stator and rotor circuits. A simplified model that neglects stator transients known as the third order model has been used in [9]. This model represents the main dynamic behavior of the induction machine and the computational complexity is low; in this case, the two derivatives in the stator voltage of (3.9) are set to zero.

The fixed speed wind generator consumes a large amount of reactive power, especially during the magnetization of the machine. In the case of voltage dip, the asynchronous generator is demagnetized and the consumption of reactive power increases notably, leading to a decrease in the voltage. Under these circumstances, the grid voltage is hardly recovered. Capacitor banks are used to reduce the demand of reactive power.

Reactive power support of fixed-speed wind farms can be achieved by adding an external var source, such as SVC or STATCOM, as shown in [10]. In this case, the reactive power capability is constrained to the externally added FACTS capability.

3.4 Doubly Fed Induction Generators

The most common variable-speed wind turbines are the Doubly Fed Induction Generator (DFIG), which offers high efficiency over a wide range of wind speeds as well as the ability to supply power at a constant voltage and frequency while the rotor speed varies. This technology consists of a wound rotor induction generator and a back-to-back power converter placed into the rotor of the machine while the stator is directly connected to the grid (Fig. 3.2). The power converter allows for the machine to be controlled between sub-synchronous speed and super-synchronous speed (a speed higher than the synchronous speed), usually, a variation from −40 % to +30 % of synchronous speed is chosen [11]. This converter capacity is designed to handle 20–30 % of the machine rate, which is beneficial both economically and technically.

3.4.1 Steady State Model of the DFIG

The steady state stator and rotor equations, considering a symmetrical three-phase system could be written following the complex phasor notation [12]:

$$\underline{U}_S = (R_S + j\omega_1 L_S)\underline{I}_S + j\omega_1 L_0 \underline{I}_R \tag{3.13}$$

$$\frac{\underline{U}_R}{s} = \left(\frac{R_R}{s} + j\omega_1 L_R\right)\underline{I}_R + j\omega_1 L_0 \underline{I}_S \tag{3.14}$$

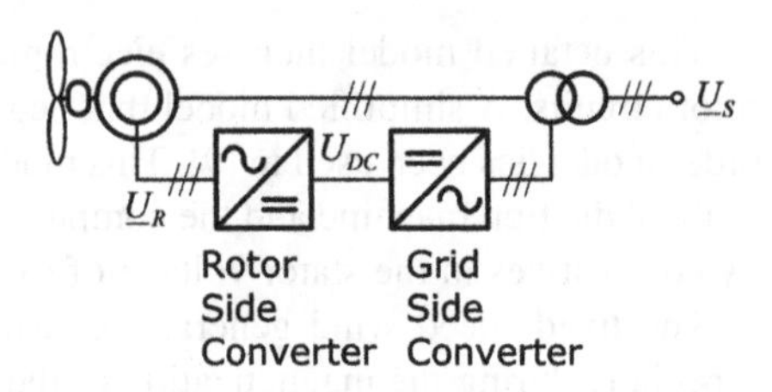

Fig. 3.2 Double fed induction generator

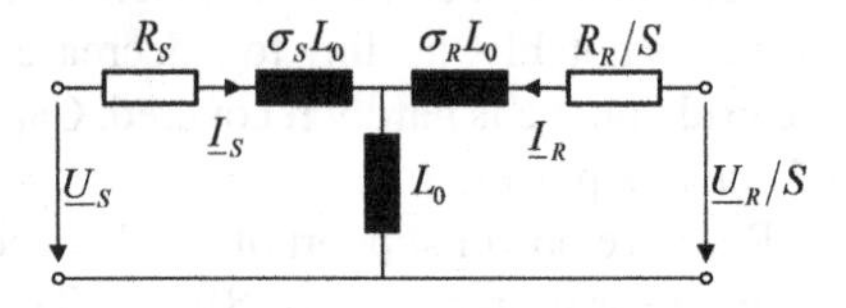

Fig. 3.3 Steady state equivalent circuit of DFIG

Introducing the stator and rotor leakage factors, the stator and rotor self inductances would be defined as:

$$L_S = (1 + \sigma_S)L_o \tag{3.15}$$

$$L_R = (1 + \sigma_R)L_o \tag{3.16}$$

And so that the mathematical model becomes:

$$\underline{U}_S = (R_S + j\omega_1\sigma_S L_o)\underline{I}_S + j\omega_1 L_o(\underline{I}_S + \underline{I}_R) \tag{3.17}$$

$$\frac{\underline{U}_R}{s} = \left(\frac{R_R}{s} + j\omega_1\sigma_R L_o\right)\underline{I}_R + j\omega_1 L_o(\underline{I}_S + \underline{I}_R) \tag{3.18}$$

And it is graphically represented in Fig. 3.3. Neglecting stator and rotor resistances results in:

$$\underline{U}_S = j\omega_1\sigma_S L_o\underline{I}_S + j\omega_1 L_o(\underline{I}_S + \underline{I}_R) \tag{3.19}$$

$$\frac{\underline{U}_R}{s} = j\omega_1\sigma_R L_o\underline{I}_R + j\omega_1 L_o(\underline{I}_S + \underline{I}_R) \tag{3.20}$$

Defining the stator magnetizing current $\underline{I}_{mS}$ as an extended magnetizing current phasor responsible for the stator flux including stator leakage [12]:

$$\underline{I}_{mS} = (1 + \sigma_S)\underline{I}_S + \underline{I}_R \tag{3.21}$$

Phasor frames are widely used in Field Oriented Control in order to simplify the equations. Using the $\underline{I}_{mS}$ phasor as the d-q reference, as shown in Fig. 3.4, the stator voltage Eq. 3.19 may be written in the form:

$$\underline{U}_S = j\omega_1 L_o\underline{I}_{mS} \tag{3.22}$$

Fig. 3.4 Phasor I_{mS} used as $d-q$ frame reference

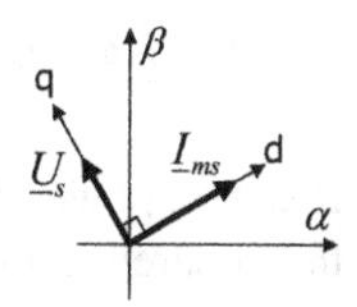

It should be noted that $\underline{I}_{mS}$ is equal to its module I_{mS}:

$$\underline{I}_{mS} = |I_{mS}| = I_{mS} \tag{3.23}$$

Hence, the stator voltage phasor is purely imaginary and aligned with the q-axis

$$\underline{U}_S = j\omega_1 L_o I_{mS} \tag{3.24}$$

3.4.2 Active Power Delivery to the Grid

Active power delivered by the DFIG stator could be deduced from the real component of the apparent power at stator terminals.

$$P_S = -\text{Re}\{\underline{U}_S \underline{I}_S^*\} \tag{3.25}$$

And so that by substituting $\underline{U}_S$ by (3.24) and $\underline{I}_S$ by its $d-q$ components: $\underline{I}_S = I_{Sd} + jI_{Sq}$, the stator active power results in:

$$P_S = -\omega_1 L_o I_{mS} I_{Sq} = -U_S I_{Sq} \tag{3.26}$$

In the same way, active power delivered by the rotor may be obtained from the real component of the apparent power at rotor terminals:

$$P_R = -\text{Re}\{\underline{U}_R \underline{I}_R^*\} \tag{3.27}$$

And so that by substituting $\underline{U}_R$ by (3.20) and $\underline{I}_R$ by (3.21), the expression for the active power delivered by the rotor becomes:

$$P_R = s\omega_1 L_o I_{mS} I_{Sq} = sU_S I_{Sq} \tag{3.28}$$

Note that rotor power P_R is a fraction of the stator power P_S depending on the slip s:

$$P_R = -sP_S \tag{3.29}$$

In this way, the total active power delivery to the grid is expressed thus:

$$P = P_S + P_R = U_S I_{Sq}(s-1) \tag{3.30}$$

3.4.3 Reactive Power Delivery to the Grid

Reactive power delivered by the stator could be obtained from the imaginary part of the apparent power as shown in:

$$Q_S = -\mathrm{Im}\{\underline{U}_S \underline{I}_S^*\} = -U_S I_{Sd} \tag{3.31}$$

Reactive power delivered by the rotor will depend on the DFIG control strategy defined in the Grid Side Converter. It is usually designed to maintain the DC voltage between the two converters at a constant operational level as well as to flow the rotor active power to the grid $P_{GSC} = P_R$. Because of the fact that the reactive power reference is usually kept to zero, the only suitable way to regulate the total power factor of the machine at the connection point PCC (Point of Common Coupling) would be by controlling the stator reactive power by (3.31).

However the grid side converter could be used to control the reactive power and to improve the total reactive power capability of the wind turbine [13, 14] too. This potential usage is a very important aspect since it may be quite useful to system operators in order to perform a coordinated reactive power management in the whole power network. The proposed methodology could be applied in the available converter designs not being necessary to perform any physical modification to the current DFIG commercial converters. And therefore the grid side converter could be treated as a reactive power source dynamically controlled.

Total reactive power injected to the grid will be composed by the superposition of both, the reactive power injected by the induction machine (stator) and the reactive power injection from the Grid Side Converter (GSC).

$$Q = Q_S + Q_{GSC} \tag{3.32}$$

3.4.4 DFIG Limitations on Deliverable Power

The maximum power capability of the DFIG wind turbine is defined by the following constraints:

- Stator Side: Both maximum voltage and current are bounded by stator machine rating.

$$I_S < I_{S,max} \tag{3.33}$$

$$U_S < U_{S,max} \tag{3.34}$$

Voltage at the PCC, U_S is allowed to range between 10% U_{PCC} according to the voltage grid regulations.

- Rotor Side Converter: Both maximum voltage and current are restricted by rotor machine rating and by Rotor Side Converter (RSC) rating:

$$I_R < I_{R,max} \tag{3.35}$$

$$U_R < U_{R,max} \tag{3.36}$$

- Grid Side Converter: Apparent GSC power should not exceed its rated or nominal value.

$$S_{GSC} < S_{GSC,nominal} \tag{3.37}$$

- Slip speed range may oscillate between the maximum and minimum values $\{s_{min}, s_{max}\}$ that are specified by the manufacturer. Slip operation point of the DFIG is obtained from (3.6).

Keeping in mind the main purpose of getting a better understanding of the implications related to each constraint, a PQ diagram will be obtained and analyzed individually for each of them in following sections.

3.4.4.1 Rotor Current Limit

The rotor current phasor could be characterized by the maximum complex modulus where the phase, φ, is swept over the operating range.

$$\underline{I}_R = I_{R,max} e^{j\varphi} = I_{R,max}(\cos\varphi + j\sin\varphi) \tag{3.38}$$

For each working condition the steady-state of the DFIG machine is obtained and a PQ locus for active and reactive powers can be drawn.

Stator current could be expressed in the $d-q$ reference frame, according to:

$$\underline{I}_S = I_{Sd} + jI_{Sq} \tag{3.39}$$

Using (3.31), the $d-q$ components are:

$$I_{Sd} = \frac{I_{msd} - I_{R,max}\cos\varphi}{(1+\sigma_S)} \tag{3.40}$$

$$I_{Sq} = \frac{-I_{R,max}}{(1+\sigma_S)}\sin\varphi \tag{3.41}$$

Substituting (3.40) and (3.41) in (3.30) and (3.32), the resulting expressions for active and stator reactive power will be formulated as follows:

$$P = -\frac{U_S I_{R,max}}{1+\sigma_S}(s-1)\sin\varphi \tag{3.42}$$

$$Q_S = \frac{U_S I_{R,max}}{1+\sigma_S}\cos\varphi - \frac{U_S^2}{\omega_1 L_o(1+\sigma_S)} \tag{3.43}$$

Resulting in the power locus PQ diagram

$$\left[\frac{P}{s_o^{I_r}(s-1)}\right]^2 + \left[\frac{Q_S - Q_o^{I_r}}{s_o^{I_r}}\right]^2 = 1 \tag{3.44}$$

where $s_o^{I_r}$, $Q_o^{I_r}$ are defined as:

$$s_o^{I_r} = \frac{U_S L_o}{L_S} I_{R,max} \tag{3.45}$$

$$Q_o^{I_r} = \frac{U_S^2}{\omega_1 L_S} \tag{3.46}$$

The resulting PQ diagram considering the rotor current constraint is an ellipse centered in $[0, Q_o^{I_r}]$ with semi-axes $[s_o^{I_r}(s-1), s_o^{I_r}]$, as shown in Fig. 3.5. Moreover, when slip is zero (synchronous speed), the PQ diagram becomes a circumference with center in $[0, Q_o^{I_r}]$ and radius $s_o^{I_r}$.

3.4.4.2 Rotor Voltage Limit

Rearranging the rotor voltage phasor (3.20) using $\underline{I_S}$ from (3.21) and $\underline{U_S}$ from (3.24) results in:

$$\underline{U_R} = js[(1+\sigma_R)U_S + \underline{I_S}\alpha] \tag{3.47}$$

where α is defined as:

$$\alpha = \omega_1 L_o[1 - (1+\sigma_S)(1+\sigma_R)] \tag{3.48}$$

By considering the rotor voltage phasor $\underline{U_R}$ as,

$$\underline{U_R} = U_{R,max}e^{j\varphi} = U_{R,max}(\cos\varphi + j\sin\varphi) \tag{3.49}$$

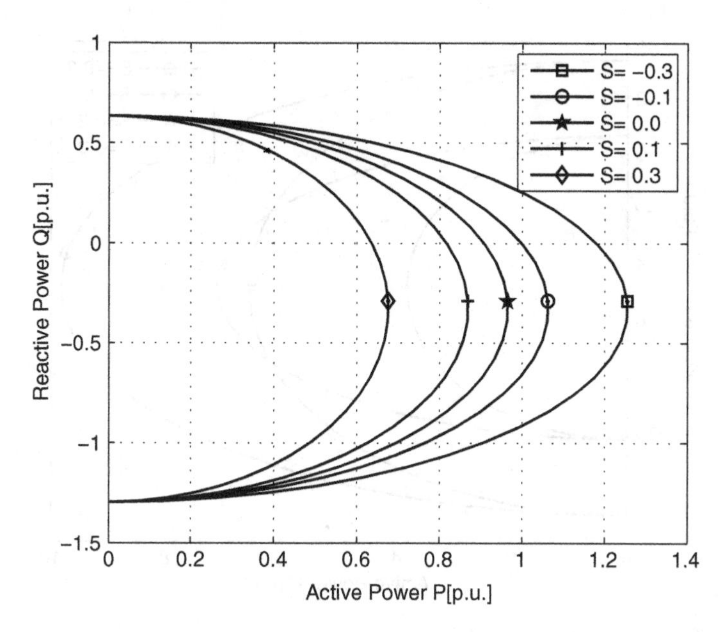

Fig. 3.5 PQ locus for rotor current $I_{R,max}$

and using (3.47), the $d - q$ components are:

$$U_{Rd} = -\alpha s I_{Sq} \tag{3.50}$$

$$U_{Rq} = s(\alpha I_{Sd} + (1 + \sigma_R)U_S) \tag{3.51}$$

Active power and stator reactive power could be rewritten thus:

$$P = \frac{U_S(1-s)}{s\alpha} U_{R,max} \cos\varphi \tag{3.52}$$

$$Q_s = \frac{1+\sigma_R}{\alpha} U_s^2 - \frac{U_S}{s\alpha} U_{R,max} \sin\varphi \tag{3.53}$$

Resulting in the power locus PQ diagram:

$$\left[\frac{P}{S_O^{U_r}(S-1)}\right]^2 + \left[\frac{Q_s - Q_O^{U_r}}{S_O^{U_r}}\right]^2 = 1 \tag{3.54}$$

where $s_o^{U_r}$, $Q_o^{U_r}$ are defined:

$$s_o^{U_r} = \frac{U_S}{s\alpha} U_{R,max} \tag{3.55}$$

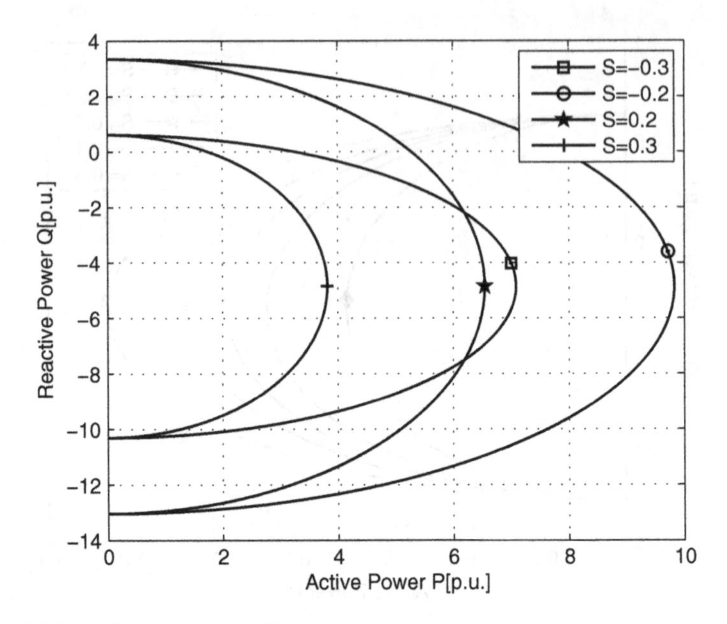

Fig. 3.6 PQ locus for rotor voltage $U_{R,max}$

$$Q_o^{U_r} = \frac{1+\sigma_R}{\alpha} U_S^2 \tag{3.56}$$

The PQ diagram is an ellipse centered in $[0, Q_o^{U_r}]$ with semi-axes $[s_o^{U_r}(s-1), s_o^{U_r}]$ as shown in Fig. 3.6. When slip is zero, the PQ diagram becomes a circumference with center in $[0, Q_o^{U_r}]$ and radius $s_o^{U_r}$.

3.4.4.3 Stator Current Limit

Active power and stator reactive power could be rewritten as:

$$P = U_S(1-s)I_{S,max}\sin\varphi \tag{3.57}$$

$$Q_S = -U_S I_{S,max}\cos\varphi \tag{3.58}$$

Resulting in the power locus PQ diagram:

$$\left[\frac{P}{s_o^{I_S}(s-1)}\right]^2 + \left[\frac{Q_S}{s_o^{I_S}}\right]^2 = 1 \tag{3.59}$$

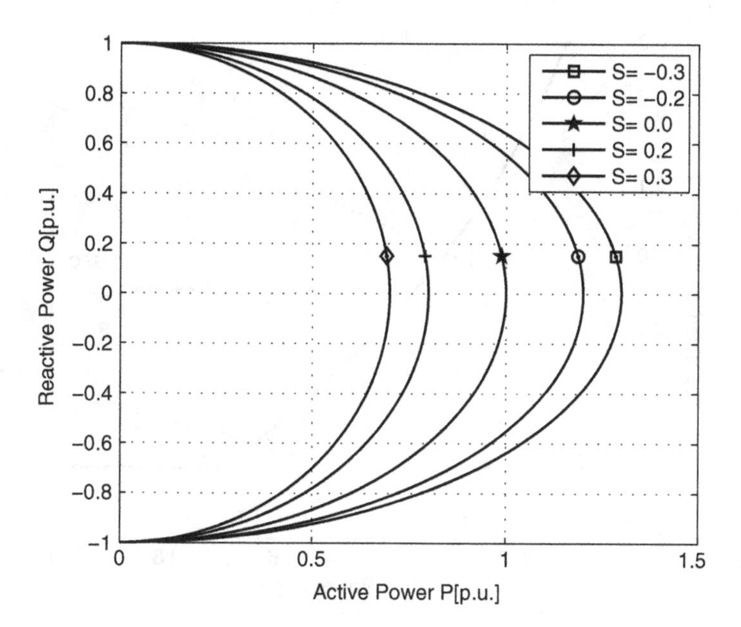

Fig. 3.7 PQ locus for stator current $I_{S,max}$

where $s_o^{I_S}$ is defined as:

$$s_o^{I_S} = U_S I_{S,max} \tag{3.60}$$

The PQ diagram is an ellipse centered in the origin $[0,0]$ and semi-axes $[s_o^{I_S}(s-1), s_o^{I_S}]$ as shown in Fig. 3.7. When slip is zero, the PQ diagram becomes a circumference with radius $s_o^{I_S}$.

3.4.4.4 Grid Side Converter Limit

The Grid Side Converter rating S_{GSC} will define the limits of the reactive power injected to the grid according to the following expression graphically shown in Fig. 3.8:

$$Q_{GSC} = \pm\sqrt{S_{GSC}^2 - P_R^2} \tag{3.61}$$

3.4.5 Maximum DFIG Capability

Taking into account all previous constraints, the resulting capability curve is shown in Fig. 3.9. It should be highlighted that when the machine is delivering reactive

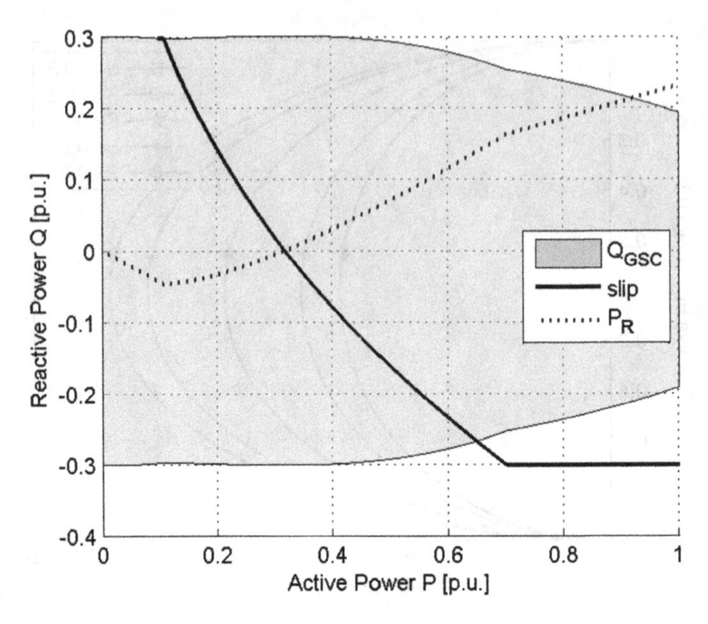

Fig. 3.8 PQ locus for grid side converter

power to the grid, rotor current is the critical constrain; whereas when the machine is consuming reactive power, stator current works as the limit factor.

In the same Figure it is to be noted two regions: the inner area corresponds to the PQ diagram without considering the reactive power injection from the GSC, and the outer area considers the reactive power capability from the GSC (Extended region in Fig. 3.9).

This extended capability could be very useful to system operators not only by performing voltage stability or contingency analysis, but also in every situation where power reactive reserves are critical factors for keeping stable the network.

3.5 Direct Drive Generators

During the last years, direct drive wind generators of big size have began to be installed by employing an electric power converter. According to report [15], in 2011 direct drive wind generators supplied 21.2 % of world wind energy production. Furthermore, they present great growth expectation regarding their employment at big offshore wind farms.

One of the main advantages of this type of generators is that they allow a total control of wind generator since not only reactive power, but also generator's

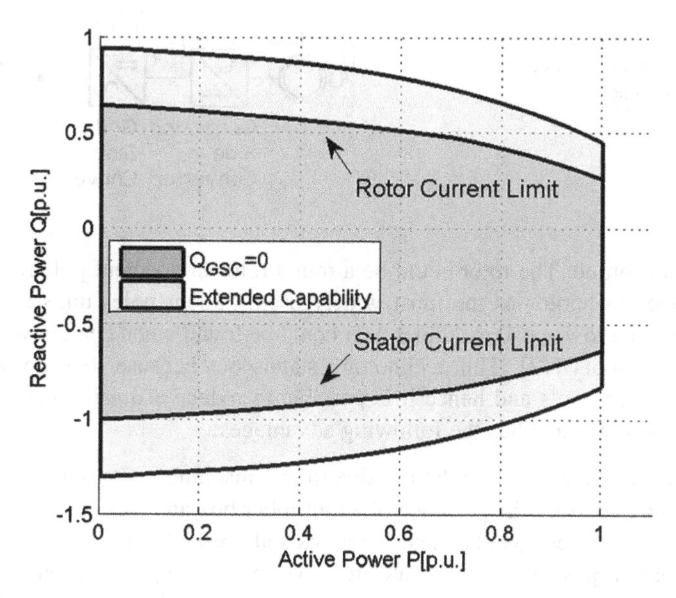

Fig. 3.9 PQ locus capability

mechanical torque could be controlled in a very detailed form throughout all the performance range. This mentioned control capacity has made them to become one of the best adapted technologies to the requirements appointed by the Transmission System Operators (TSO) at grid codes.

On the other hand, since these wind generators are connected directly to the network by means of power converters there is no need of installing additional reactive power compensation equipments, as for instance FACTS, and thus it is possible to reduce the required infrastructure expenses to better fulfill those requirements appointed by network operators.

The direct-drive generator can be classified into:

- Induction Generator: This technology corresponds to induction machines equipped with brushless squirrel cage rotor connected to the network by means of electric power converters. This technology is revealed as being robust, very effective at partial load as well as incurring low costs of the electric machine. As drawback, it must be pointed out the need of incorporating large-diameter wind turbines (which entail considerable installations problems too), it presents high prices and converters must be sized in order to control the total power developed by the machine as well as the required power to the magnetization process [16].
- Wound rotor synchronous generator: In the present case, the electric machine is a wound rotor synchronous generator able to regulate its own magnetization level by using an auto-excitation circuit and therefore it exerts control over the

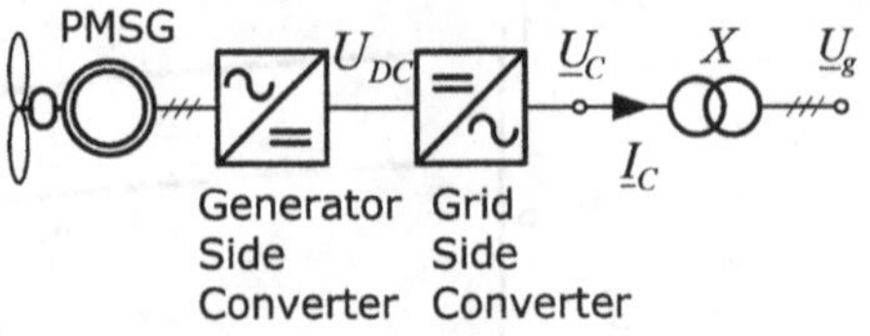

Fig. 3.10 Equivalent circuit of the direct driven variable speed wind turbine

generator torque. The rotor could be a round rotor or a salient pole rotor being multipoles technologies the most widely used. Salient poles units are mainly employed by low-speed machines as it could be found when dealing with large-scale generators [17]. This technology stands out because they do not need permanent magnets and hence it is possible to reduce acquisition and maintenance costs [18] and has the following advantages:

- Limited mechanical oscillations due to low machine's rotation.
- It is no necessary the presence of a multiplier box and therefore it is possible to reduce prices, maintenance problems and noise effect.
- It makes possible to regulate reactive power depending on network's requirements.
- It allows exerting an independent control of active and reactive power.

Among the Drawbacks:

- Large size and heavy machines.
- It is required to supply a high electric torque under low-speed conditions.
- High converter's cost.

- Permanent synchronous magnet generator: Synchronous generators made of permanent magnets are synchronous units being magnetized from permanents magnets located at rotor instead of being forced to use a continuous excitation circuit. This configuration allows reducing electric losses and improving machine's thermal properties [19]. Synchronous generators are connected to the electronic power converter, which for its part is made of a rectifier (converter AC/DC), a chopper (converter DC/DC) and an inverter (converter DC/AC) (Fig. 3.10). A first classification taking into account permanent magnets' location could be if they are surface-mounted or inserted into the rotor. As long as magnetic field flow is regarded, there are different technologies as indicated at [20].
  - Radial-flux permanent synchronous magnet units: This technology is the most widespread among permanent synchronous magnet units. It consists of a synchronous machine with radial flux in the air gap and longitudinal flux at stator. Magnets are located inside the rotor.
  - Stator axial-flux permanent synchronous magnet units: They are also called units of transversal flow and the properties which characterize

them are that rotor's flux moves in a transversal way whereas stator's flux in an axial way. Magnets are situated inside the rotor and because of air gap, the flow is impelled to follow a radial movement. It is equipped with an externally located rotor and a stator located inside the device. Magnets are situated inside the rotor forming two rows for each phase, with the main goal of allowing flow circulation through the stator.

- Rotor axial-flux permanent magnet generator: They are made of two externally situated disks forming the rotor. Permanent magnets are located on the surface of rotor disks and therefore flux moves parallel to rotation axis.

Permanent synchronous magnet units are mainly multipoles counting with a high diameter of turbine's rotor and their main advantages are:

- Lightweight design.
- High efficiency relation power/weight.
- Reduced operating and maintenance costs.
- Negligible rotor's losses.
- Higher reliability due to the lack of mechanical components (slip rings, brushes).
- There is no need of extra power to supply the winding magnetic excitation field.

As main drawbacks:

- High initial inversion due to permanent magnet's costs.
- Risk of machine's demagnetization under circumstances of high temperature, overload or short circuits.
- The use of diodes by the rectifier reduces control possibilities.
- Handling difficulties during construction process.

3.5.1 Power Capability of Direct Driven Wind Generators

Direct drive wind generators present as main characteristic that electric machine's stator is connected directly to the network by means of a power converter. This converter consists of a rectifier, a DC/DC converter and an inverter with Pulse-Width Modulation (PWM) or Space Vector Modulation (SVM), which for its part could be made of three-phase Insulated Gate Bipolar Transistors (IGBT).

The interchange of machine's reactive power with the network is defined by those control techniques applied to the power converters where converter's operating limits are considered, it should to be highlighted that the power capability of the Direct Driven wind turbine is not limited by the electrical machine's properties nor machine's characteristics. Acting this way, the direct driven variable speed wind turbine is capable to supply a whole regulation of reactive power and active power regulation under different network's conditions or when facing a set point change commanded by the central wind farm control [21].

3.5.1.1 Direct Drive Limitation on Deliverable Power

The maximum power capability of the Full Power Converter is defined by the following constraints:

- Converter Current that should not exceed a maximum value $I_{c,max}$:

$$I_c < I_{c,max} \tag{3.62}$$

- Converter Voltage that should not be superior to a maximum value $U_{c,max}$:

$$U_C < U_{c,max} \tag{3.63}$$

- DC link Voltage that should not exceed a maximum limit $U_{DC,max}$:

$$U_{DC} < U_{DC,max} \tag{3.64}$$

3.5.1.2 Current Converter Limit

For each working condition the power delivered to the grid is:

$$\underline{S} = \underline{U}_g \underline{I}_c^* \tag{3.65}$$

where the voltage at the PCC ($\underline{U}_g$) is bounded between maximum and minimum values.

$$\underline{U}_{g,min} < \underline{U}_g < \underline{U}_{g,max} \tag{3.66}$$

The grid side converter delivers power to the grid according to the following equation:

$$P^2 + Q^2 = (U_g I_c)^2 \tag{3.67}$$

where the converter current is:

$$I_c = \frac{\sqrt{P^2 + Q^2}}{U_g} \tag{3.68}$$

and takes the maximum value at the minimum grid voltage:

$$I_{c,max} = \frac{\sqrt{P^2 + Q^2}}{U_{g,min}} \tag{3.69}$$

3.5.1.3 Voltage Converter Limit

The converter is connected to the grid though a transformer characterized by a pure reactance (X) and the voltage at the converter output ($\underline{U}_c$) is given by:

$$\underline{U}_C = jX\underline{I}_c + \underline{U}_g \tag{3.70}$$

The converter current I_c is:

$$\underline{I}_c = \frac{\underline{U}_C - \underline{U}_g}{jX} \tag{3.71}$$

The complex power at the grid side converter output may be written as

$$\underline{S} = \underline{U}_g\underline{I}_c^* = \underline{U}_g\left(\frac{\underline{U}_C - \underline{U}_g}{jX}\right)^* \tag{3.72}$$

where real and imaginary parts are of form

$$P + jQ = \underline{U}_g\left(\frac{\underline{U}_C - \underline{U}_g}{jX}\right)^* \tag{3.73}$$

If the converter voltage is expressed in the $d - q$ reference frame, according to:

$$\underline{U}_c = U_C \exp(j\varphi) \tag{3.74}$$

Substituting (3.74) in (3.73), real and imaginary parts of the power can be expressed as:

$$P = \frac{U_g U_C}{X}\cos(\pi 2 - \varphi) \tag{3.75}$$

$$Q = \frac{U_g U_C \sin(\pi 2 - \varphi) - U_g^2}{X} \tag{3.76}$$

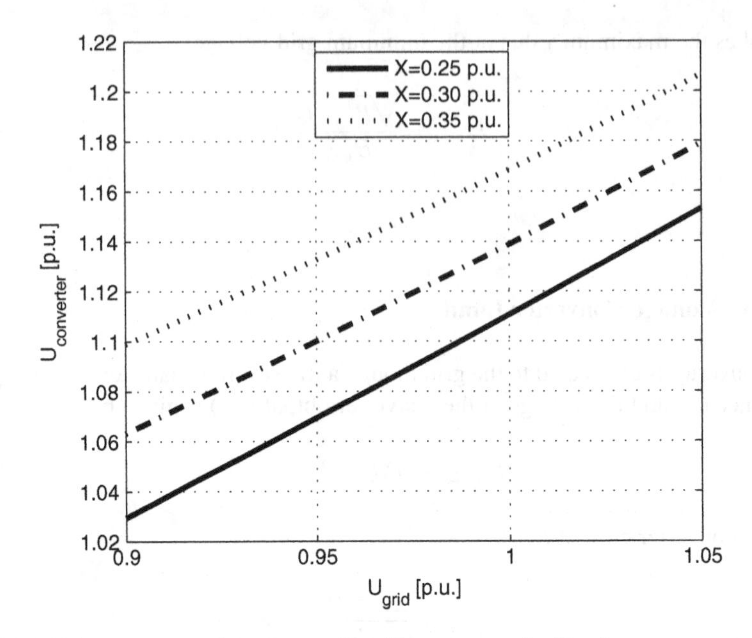

Fig. 3.11 Voltage converter dependency with grid reactance and grid voltage

Resulting in the power locus PQ diagram:

$$P^2 + \left(Q + \frac{U_g^2}{X}\right)^2 = \left(\frac{U_g U_C}{X}\right)^2 \tag{3.77}$$

Grid Code requirements specify that wind turbines shall offer reactive power capability for each operational point, even for the maximum active power. Consequently, this requirement can be characterized by a power factor ($cos(\phi_R)$) at the rated active power (P_R).

By using U_C from (3.77) at the rated value, voltage at the converter output is given by:

$$U_C = \frac{X}{U_g}\sqrt{P_R^2 + \left(P_R\,\tan(\phi_R) + \frac{U_g^2}{X}\right)^2} \tag{3.78}$$

Figure 3.11 depicts the U_C dependency with the grid reactance and voltage. The admissible range for the grid reactance (X) and for the grid voltage (U_g) has been varied. It can be noted how a decrease of the grid voltage produces a decrease in the converter voltage, in respect to the grid reactance variation. In the same figure it is shown that the required converter voltage is higher for more inductive grids. The maximum value $U_{C,max}$ is found at X_{max} and $U_{g,max}$.

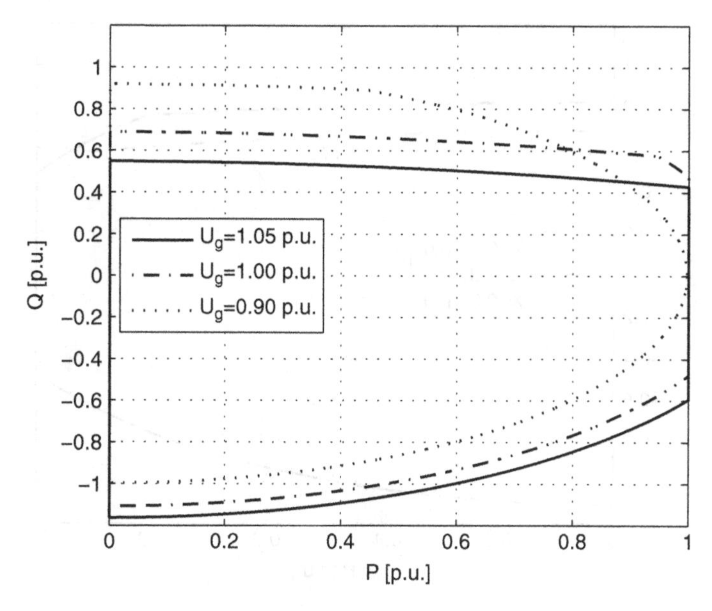

Fig. 3.12 Grid voltage influence on the PQ capability

3.5.1.4 DC Link Voltage Limit

In order to obtain the maximum converter voltage $U_{C,max}$, the DC link shall be able to work at $U_{DC,max} = \sqrt{2}U_{C,max}$, considering the PWM or SVM modulation strategies.

3.5.2 PQ Capability

Considering the previous constraints the resulting $P - Q$ capability is restricted to the common regions between the current converter limit (3.67) and the voltage converter limit (3.77) for a required power factor ($cos(\phi_R)$).

In order to obtain the $P - Q$ capability dependency on the grid voltage, the grid coupling reactance and the required power factor, a series of tests have been performed. In each test only one parameter is swept through the operating region keeping the rest parameters fixed to their rated values.

3.5.2.1 Grid Voltage Influence

Figure 3.12 shows the PQ locus where the grid voltage varies from 0.90 to 1.05 p.u. In this case the reactive grid value and the rated power factor has been fixed in all

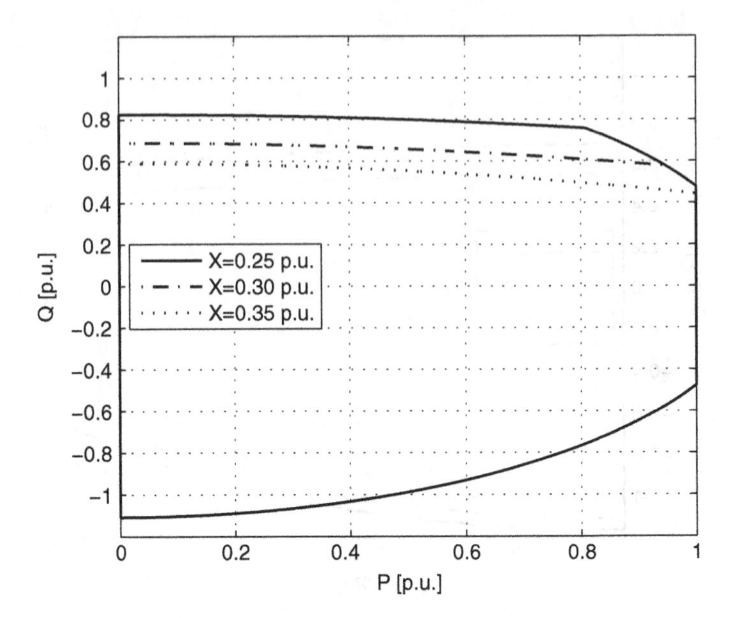

Fig. 3.13 Grid reactance influence on the PQ capability

simulations. From these PQ curve, it can be noted that the PQ capability is bounded by the current converter limit (upper limit) and by the voltage converter limit (lower limit). If the direct drive wind generator is properly designed, it is possible to offer reactive power support even for the rated active power, although, its reactive power capability relies on the grid voltage, if the grid voltage reduces the reactive power capability is lower.

3.5.2.2 Grid Reactance Influence

In this test grid copling reactance has been varied from 0.25 to 0.35 p.u keeping the rest of the parameters (grid voltage and power factor) fixed in their rated values. Figure 3.13 shows that the reactive power capability of the full power converter is higher when it is connected to lower inductive networks

3.5.2.3 Power Factor Influence

Figure 3.14 shows the influence that the power factor has on the PQ capability of the full power converter. It is noted how the wind turbine is able to produce reactive power when the reference power factor is keep to1.00. In cases where the machine is required to work under leading-lagging power factor 0.95, it can be seen that even at rated operation (P_{rated} = 1.0 [p.u]) the machine is able to offer reactive power capability to the grid.

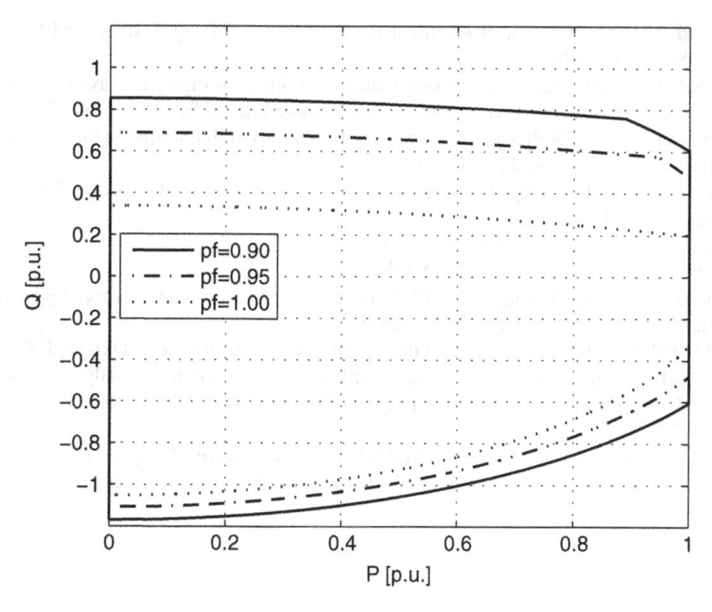

Fig. 3.14 Power factor influence on the PQ capability

References

1. Braun M (2008) Reactive power supply by distributed generators. In: Proceedings of IEEE power and energy society general meeting, Pittsburgh, PA, USA, pp 1–8
2. Hedayati H, Nabaviniaki SA, Akbarimajd A (2008) A method for placement of DG units in distribution networks. IEEE Trans Power Deliv 23(3):1620–1628
3. Vijayan P, Sarkar S, Ajjarapu V (2009) A novel voltage stability assessment tool to incorporate wind variability. In: Power energy society general meeting. PES '09, IEEE, Calgary, AB, Canada, pp 1–8
4. Sangsarawut P, Oonsivilai A, Kulworawanichpong T (2010) Optimal reactive power planning of doubly fed induction generators using genetic algorithms. In: Proceedings of the 5thIASME/WSEAS international conference on energy, Cambridge, UK, pp 278–282
5. Ping-Kwan Keung, Yuriy Kazachkov, Senthil J (2009) Generic models of wind turbines for power system stability studies. In: Proceedings of the 8th international conference on advances in power system control, operation and management (APSCOM 2009), London, UK, pp 1–6
6. Vilar Moreno C, Amaris Duarte H, Usaola Garcia J (2002) Propagation of flicker in electric power networks due to wind energy conversions systems. IEEE Trans Energ Convers 17:267–272
7. Freris LL (1990) Wind energy conversion systems. Prentice Hall, New York
8. Kundur P (1994) Power system stability and control. McGraw-Hill, New York
9. Krause PC, Nozari F, Skvarenina TL, Olive DW (1979) The theory of neglecting stator transients. IEEE Trans Power Apparatus Syst PAS-98(1):141–148
10. Slepchenkov MN, Smedley KM, Jun W (2011) Hexagram-converter-based STATCOM for voltage support in fixed-speed wind turbine generation systems. IEEE Trans Ind Electron 4:1120–1131
11. Ackermann T (2005) Wind power in power systems. Wiley, Hoboken

12. Leonhard W (1990) Control of electrical drives (Electric Energy Systems and Engineering Series). Springer, Berlin
13. Ullah NR, Thiringer T (2008) Improving voltage stability by utilizing reactive power injection capability of variable speed wind turbines. Int J Power Energy Syst 28(3):289–229
14. Bharat S, Singh SN (2009) Reactive capability limitations of doubly-fed induction generators. Electric Power Comp Syst 2009:427–440
15. International Wind Energy Development (2011) World Market Update 2011 Forecast 2012–2016. BTM Consultant ApS
16. Technical characteristics Siemens Bonus SWT-3.6-120
17. Enercon: Product Overview E-126/7.5 MW. Apr 2012
18. Baroudi JA, Dinavahi V, Knight AM (2007) A review of power converter topologies for wind generators. Renew Energy 32(14):2369–2385
19. Vas P (1992) Electrical machines and drives. Oxford University Press, Oxford, UK
20. Saleh SA, Khan MASK, Rahman MA (2011) Steady-state performance analysis and modeling of directly driven interior permanent magnet wind generators. IET Renew Power Gener 5 (2):137–147
21. Ullah NR, Bhattacharya K, Thiringer T (2009) Wind farms as reactive power ancillary service providers – technical and economic issues. IEEE Trans Energ Convers 24:661–672

Chapter 4
Reactive Power Optimization

Every optimization problem employed to study electric power systems consists of an objective function and a group of constraints to be observed by this function that concurrently define the problem itself. The main constraints associated with the problem of reactive power planning are related to load flow equations. These problems are widely known as optimum load flow. Current operating conditions of reactive systems imply the necessity of redefining the reactive power planning problem, though bearing in mind its performance under normal conditions and faults or disturbances. The employment of these constraints by the planning of reactive power leads to a second optimization model known as optimum load flow with security constraints. Finally, more recent studies have raised the possibility of including voltage stability into the objective function, as well as into the problem constraints of reactive power planning, to maximize the voltage stability margin. Thus, a third optimization problem is derived, namely optimum load flow with security and voltage stability constraints. This optimization method aims to guarantee the existence of voltage stability margins in the system under faults and disturbances.

4.1 Introduction to the Optimum Load Flow

The first optimization method employed when studying electric power systems was the optimal power flow, [1, 2], where the problem can be defined by 2n unknown factors (control variables) and 2n equations, "n" being the number of buses in the system. Usually, a conventional power flow reaches a feasible system's solution that takes into account only the technical restrictions. By acting in this manner, it set apart the fulfillment of any possible objective function that could be additionally included, for example, costs associated with the system. To find out a suitable solution, an optimum load flow was developed to obtain feasible solutions of the problem by minimizing an objective function, for instance, cost reduction. In this case, and because of the elements employed in the power system, the value of the 4n known variables are not steady state values but could vary within a range. Active

H. Amaris et al., *Reactive Power Management of Power Networks with Wind Generation*, Lecture Notes in Energy 5, DOI 10.1007/978-1-4471-4667-4_4,

and reactive power injected by generators units in the system are considered to be within an allowed range. This implies that the system's resolution establishes a range of possible solutions inside which the final solution to the problem it is found. Therefore, the optimum load flow could be considered as a non-linear optimization problem subject to static constraints.

4.2 Formulation

The mathematical formulation corresponds to a general problem of numeric optimization that is subject to constraints:

$$Minimize \quad f(u,x) \tag{4.1}$$

subject to constraints:

$$g(u,x) = 0 \tag{4.2}$$

$$h(u,x) \geq 0 \tag{4.3}$$

Where:

u Represents system control variables

x Stands for the group of state or system-dependent variables

$f(u,x)$ Being the objective function

$g(u,x)$ Represents equality constrains; for example, load flow equations

$h(u,x)$ Includes inequality constraints; for example, fixed limits of control variables and operating limits of system's units

The most widely used control variables are:

- Power output of generators;
- Voltage regulation of system's buses;
- Transformer Tap regulation;
- Phase change converters;
- Devices of reactive power in parallel connection;
- Loads to be connected under special system's operating conditions;

As far as state variables are concerned, the most representative are:

- Voltage magnitude at system's buses;
- Voltage phase at system's buses;
- Power transmission through the lines;

4.3 Constraints

The main constraints applied to the optimum load flow could be represented by (4.2) and (4.3); that is, by using a group of equality constrains such as the ones corresponding to conventional load flow equations, as well as an inequality constraints such as the physical limits of control u system's variables or the physical limits of the x state.

- Equality constraints of load flow can be described by:

$$P_{gi} - P_{di} - P(U,\theta) = 0 \tag{4.4}$$

$$Q_{gi} + Q_{Ci} - Q_{di} - Q(U,\theta) = 0 \tag{4.5}$$

- Inequality constraints consider:
  - Control variables' limits:

$$P_{gi,min} \le P_{g_i} \le P_{gi,max} \tag{4.6}$$

$$U_{gi,min} \le U_{g_i} \le U_{gi,max} \tag{4.7}$$

$$T_{l,min} \le T_l \le T_{l,max} \tag{4.8}$$

$$Q_{Ci,min} \le Q_{Ci} \le Q_{Ci,max} \tag{4.9}$$

  - State variables' limits:

$$Q_{gi,min} \le Q_{gi} \le Q_{gi,max} \tag{4.10}$$

$$U_{i,min} \le U_i \le U_{i,max} \tag{4.11}$$

$$LF_l \le LF_{l,max} \tag{4.12}$$

  - Other limits: Restriction of power factor, etc.
    Where:

P_{gi} Active power injected by generator i^{th};
P_{li} Load Active power demand at bus i^{th};
Q_{gi} Reactive power injected by generator i^{th}
Q_{Ci} Reactive power injected by the shunt var unit connected to bus i^{th}
Q_{li} Load Reactive power demand at bus i^{th}

U_{gi} Generator Voltage at bus i^{th};
T_l Transformer tap position, l;
U_i Voltage at bus i^{th};
LF_i Power flow through line l;

4.4 Models Based on Voltage Stability Concepts

Several computational methods based on the bifurcation theory have proved to be an efficient tool for studying voltage stability. However, from a technical perspective, it is quite complicated to incorporate operating limits, and it can also be very expensive in the case of continuing methods, especially in large systems provided with multiple limits. The use of optimization methods for this kind of study presents various advantages, mainly related to the capability of dealing with constraints. Throughout the bibliography relating to reactive power management and voltage stability, it is mentioned the possibility of turning the voltage stability margin into a voltage index in order to detect those buses more sensitive to the incorporation of reactive compensation units [3, 4] into the studied system. Traditionally, the optimum load flow with security restrictions did not take into account changes affecting the voltage collapse point after the incorporation of new reactive compensation units into the system. Due to this recent necessity, a new optimization model has been developed in which different constraints are included relating to voltage stability by the study of cases associated with contingencies. This model is known as the optimum load flow with security and voltage stability constraints. References [5, 6] incorporate Voltage Stability margins (VSM) in the system constraints for two main reasons: The first one is related to the need to guarantee that the voltage levels at all buses along the system are within a certain range under normal operating conditions; The second reason is consistent with the necessity for confirming that the distance between the system's operating point after a contingency and the new voltage collapse point will be superior to a predefined margin (voltage stability margin). To ensure that the system meets security conditions every time, the optimization problem consists of two groups of constraints associated with the system's performance under normal conditions and after a contingency. By increasing the number of constraints, as well as by duplicating the variables associated with the two operating methods, the optimization model to solve the problem of reactive power planning will become remarkably more complicated.

4.5 Selection of Objective Function

The objective function concerning problems of reactive power management is usually based on a cost function, such as:

4.5.1 *Minimization of Variable Costs*

The basic objective function is the minimization of a system's variable costs associated with the incorporation of reactive power generation units into the system. There are two models representing this objective function. The first model considers that reactive power generation costs vary linearly with the reactive source size (Q_C), (4.13), [1].

$$Cost \quad per \quad hour_{model1} = C_1.Q_C \tag{4.13}$$

Where:

C_1 \$/Mvar.hour
Q_C Mvar

It is important to highlight that this model only takes care of the global sizing of generation units that are incorporated into the system, in other words, the quantity of total reactive power. Therefore, we would obtain the same cost whether we were incorporating two units of 200 Mvar to the system or just one unit of 400 Mvar.

The second model considers the incorporation of fixed costs (C_0), related to the device's life span cost, prorated with the life hours thereof (C_1 variable cost). This model is represented by (4.14) [5, 7, 8].

$$Cost \quad per \quad hour_{model2} = (C_0 + C_1.Q_C).x \tag{4.14}$$

Where:

C_0 \$/hour
C_1 \$/Mvar.hour
Q_C Mvar

x is a state variable indicating the connection or disconnection of a reactive power generation unit

Despite the fact that, the employment of the "x" binary variable represents a miniscule variation between the two cost models, it implies the employment of different resolution models. Hence the resolution methods applied by the first cost model are related to the conventional lineal or non-lineal programming; meanwhile, the problem's resolution set up by the second cost model implies the employment of integer-mix optimization techniques.

4.5.2 Minimization of Variable Costs and Active Power Losses

The incorporation of the cost incurred from active power losses in the system implies an evolution of the basic objective function. From the two models of variable costs proposed in the previous section, it is possible to obtain two new objective functions including the system's power losses [2, 9].

$$Cost \quad per \quad hour_{model3} = C_1(Q_c) + C_2(P_{loss}) \tag{4.15}$$

$$Cost \quad per \quad hour_{model4} = (C_0 + C_1 . Q_c).x + C_2(P_{loss}) \tag{4.16}$$

Where C_2 represents the cost associated with real power network losses.

There are indeed studies in which active power losses stand not only for the cost incurred during normal operation but also under abnormal operating conditions such as contingencies [10]. If this happens, the objective function is (4.17):

$$\min \quad F = C_1(Q_c) + \sum_{k=0}^{N_c} C_2(P_{loss})_k \tag{4.17}$$

where k represents the case subject of study and entails all possible contingency and post-contingency situations.

4.5.3 Minimization of Variable Costs and Fuel Cost

In some studies, [11–13], the employment of the consumed fuel cost is used as an indicator of system's operating costs. The minimisation of costs associated with losses does not always imply the reduction of a system's operating costs. However, the minimisation of operating costs does imply the reduction of costs related to losses.

In this case, by employing the two basic models of the system's cost the following objective functions represented by (4.18) and (4.19), can be shown:

$$Cost \quad per \quad hour_{model5} = (C_1(Q_c) + C_T) \tag{4.18}$$

$$Cost \quad per \quad hour_{model6} = (C_0 + C_1 . Q_c).x + C_T \tag{4.19}$$

where the fuel cost C_T results from the sum of individual units costs (4.20) represented by means of a quadratic approximation (4.21), [11].

$$C_T = \sum_{i=1}^{n} f_i(P_{gi}) \tag{4.20}$$

$$f_i(P_{gi}) = a_{oi} + a_{1i}P_{gi} + a_{2i}P_{gi}^2 \tag{4.21}$$

4.5.4 *Minimization of the Deviation from a Set Point*

An alternative to the incorporation of system's costs into the objective function is the employment of deviation indicators of the control variables from their set point value. This set point value could correspond to a normal system's operating situation, as well as to a certain operating condition. At [14] it could be found an example in which the objective function is in charge of minimizing the existing distance between voltage magnitude at each bus ith and its superior limit; in short, p,

$$p = \sum_i U_{i,max} - U_i \tag{4.22}$$

where the i sub-index stands for system's buses.

4.5.5 *Multi-objective Formulation*

The main objective of reactive power management is to supply the necessary reactive power to the system to compensate for any existing imbalance, thus allowing the system to operate in a stable way, keeping a balance resulting from the combination of costs and security.

Previous sections assemble a brief description of the state of art of objective functions concerning reactive power planning. Some of these studies implement multi-objective functions, such as the functions that combine variable costs and the system's losses and in which weighting coefficients of each objective are easily defined by turning each variable of the objective function into costs. The problem becomes nevertheless more complicated when, in the same function, several variables related to costs and system's security are employed.

Some examples are presented as follows:

- In [15] and [16], the objective function F includes cost related to the development of reactive compensation sources, system's losses and voltage deviation at buses from their set point value (4.23) and (4.24) as follows:

$$\min \quad F = (C_0 + C_1.Q_C).x + C_2(P_{loss}) \tag{4.23}$$

$$\max |U_{i,max} - U_i| \tag{4.24}$$

- In [17], the objective function is defined by (4.25) in which voltage violation is minimised, as well as reactive power violation and power losses.

$$\min \quad F = k_1 \sum (voltage \quad violations)^2 + \\ + k_2 \sum (Reactive \quad Power \quad violations)^2 + P_{loss} \tag{4.25}$$

- In [18], the objective function results from the combination of variable costs, active power losses in the system, voltage deviation at buses and power flows through lines from their set point value (4.26):

$$\min \quad F = (C_0 + C_1 Q_C)x + C_2(P_{loss}) + \rho_1 \sum_i \left(\frac{|U_i - U_{i0}|^2}{\Delta U_{imax}} \right)^2 + \\ + \rho_2 \sum_i \left(\frac{|LF_i - LF_{l,set\ point}|}{\Delta LF_{l,max}} \right) \tag{4.26}$$

where U_i,$U_{i,0}$, and $U_{i,max}$ correspond to the voltage at the bus i^{th}, reference voltage and the maximal voltage deviation of bus i^{th}, respectively. In the same way, LF_i, $LF_{l,setpoint}$ and $LF_{l,max}$ correspond to the power flow through lines under current operating conditions, the reference power flow through lines and the maximum deviation of power flow through lines, respectively.

4.6 Methods to Solve the Problem of Reactive Power Planning

The optimisation methods employed to solve the problem of reactive power planning and the load flow of power systems can be roughly classified as follows:

- Conventional methods:
  - Methods based on lineal programming.
  - Methods based on non- lineal programming.
  - Methods based on Nonlinear Mixed Integer programming.
- Advanced methods

4.6.1 *Conventional Methods*

4.6.1.1 Lineal Programming

Lineal programming methods stand out as being reliable and robust techniques to solve a wide range of optimization problems characterized by lineal objects and lineal constraints.

However, their application spectrum in the field of electric systems still has some constraints due mainly to an incorrect evaluation of the system's losses and to the possibility of being trapped inside local optimum solutions. There is indeed a large body of literature dealing with the problem of reactive power planning by means of lineal programming, as for instance [19, 20]. In [21], a combination of lineal programming and parametric programming is used to find the minimum number of reactive generation units required to solve the problem.

4.6.1.2 Non-lineal Programming

In spite of the fact that lineal programming has many practical applications in the field of power systems, most of the involved variables and constraints of the optimization process of electric networks are non-lineal. Non-lineal programming could be applied to solve any kind of optimization problem containing non-lineal objectives or constraints. In the case of constraints represented by Inequality, it is necessary to specify upper and lower limits.

Among the optimization methods employed by non-lineal programming, the following techniques are to be highlighted:

- Sequential quadratic programming.
- Methods based on an extended lagrangian.
- Methods based on a generalised gradient.
- Interior point method.

4.6.1.3 Nonlinear Mixed Integer Programming

Decomposition methods have been developed in the field of optimization of large power systems. They are able to break down the problem into several sub problems, usually two, having different characteristics as far as linearity is concerned. These methods' results manage to reduce the number of iterations needed to solve the problem, the time of required computation and memory space. Moreover, the decomposition process allows for the application of different solution methods in a separate manner for each one of the single problems, and its attractiveness lies in this fact. Decomposition methods are used to solve large electric power systems. In [22], the Benders decomposition method has been applied to solve problems of technical constraints regarding the daily programming of generation, as well as of integer-mix variables and non-lineal constraints.

4.6.2 *Advanced Methods*

4.6.2.1 Heuristic Methods

Currently, there are several combinatory analysis problems that are difficult to solve in an accurate way, as is the case in reactive power management, but nevertheless they do require a solution. Aiming to find a solution to these problems, certain mathematical techniques arose enabling one to obtain a feasible solution to the problem. In other words, solutions able to fulfill problem's constraints, although not managing to optimize the objective function. They obtain a solution within the optimum environment in a short span of time and with a low computational cost. A possible definition of these algorithms could be: "simple procedures", very often resulting from common sense, that are supposed to offer a good solution (although not necessarily the optimum one) to difficult problems in an efficient manner.

One of the main advantages of heuristic methods in comparison to the conventional optimization techniques able to obtain accurate solutions is that they are flexible enough to face problem's characteristics. Heuristic methods can work with lineal and non-lineal systems and they are able to offer more than one solution, something that in some cases happens to be a great advantage because it widens out the solution's spectrum.

Among the main disadvantages of heuristic techniques, of particular note is that it is not possible to know the quality of the obtained solution, in other words, to know whether the solution given by the algorithm is close to the optimum. However, there are techniques that allow including some sort of quotations giving information about the chosen solution. Among them, relaxation is to be highlighted. If these kind of evaluating procedures cannot be implemented, there are easier ones available to evaluate the quality of the heuristic solution; for instance, to generate randomly more than one solution and then check their proximity to the solution given by heuristic methods. In sum, it can be concluded that in the case of choosing an accurate technique to solve the problem, it must be always chosen at the expense of any other heuristic one; especially by those problems in which the economic factor is the keystone.

4.6.2.2 Simulated Annealing

Simulated annealing is a local search algorithm (meta-heuristic) capable of escaping from local minimum by making probabilistic moves. From a mathematical point of view, the simulated annealing introduced by Kirkpatrick, Gellatt and Vecchi in 1983 is a stochastic algorithm trying to minimize numeric functions of a great number of variables by taking leaps in the space, and thus shifting the final solution far away from the possible local lowest value [23]. As a result, it converges asymptotically to the optimum solution with probability one [24].

This technique is based on the cooling concept used by the metallurgic sector to obtain an ordered solid state of metals when working with metals of lower energy and avoiding any possible metastable state. This technique softens the metal by subjecting it to high temperatures and then letting it cool very slowly till the particles start taking position on their own and reach the solid's "fundamental state". To allow the material to reach thermal balance, it is necessary that the cooling process takes place slowly; if this balance is reached too quickly, the solid could reach a metastable balance instead of a fundamental one in which the particles take the form of perfect reticules and the system works at its lower energy level.

Simulated annealing applies this metallurgic technique upon the mathematic programming. Therefore, the main goal is to minimize the problem's objective function, similar to the material's energy, by employing a fictitious temperature, which is a controllable parameter of the algorithm. In practice, Metropolis algorithms are used based on Monte Carlo's techniques to simulate the material's cooling. This algorithm enables one to describe the balanced thermal performance of a material under a determined temperature by spreading a large number of transitions using a Boltszmann distribution to describe the thermal balance.

This process implemented by a simulated cooling could be summarized as follows: starting with an initial state, if this state's development is provided with a lower energy (lower value of evaluation function) than the one of current state, then the resulting state will be accepted as current state. On the contrary, if the resulting state yields an increment δE in objective function, the resulting state will be accepted only with a determined probability given by $e^{-\delta E/T}$, where T stands for temperature. This acceptance probability depending on temperature allows the attainment of each state but based upon differing probabilities due to diverse temperatures.

Therefore, simulated annealing could be considered as an iterative process of Metropolis algorithms which are performed with decreasing values of the control parameter (temperature). Conceptually, Simulated annealing entails a neighborhood searching process, in which selection criterion depends on transitions rules given by the Metropolis algorithm. In this regard, acceptance probability decreases if raising the difference between the value of evaluation's function of an initial solution and the value of the candidate. The existence of this random factor when searching allows the candidate to reduce probabilities of being trapped in a local optimum.

Main disadvantages of this method are related to the adjustment of the control parameter (temperature), which depends on the ability and knowledge of the programmer in charge of the algorithm. Furthermore, the algorithm's calculating time could reach high values and thus some parallel implementation techniques start being employed. Finally, the flexibility factor stands out among all the advantages involved in this method, promoting the problem's development and a user friendly implementation.

In the existing literature, several pieces of work concerning reactive power planning by means of simulated annealing can be found. In [25], this technique has been employed to obtain type, location and optimum sizing of reactive generation units. In [18], reactive power planning is presented as an undifferentiated multi-objective problem observing constraints and being solved by a two-phase algorithm based on an extended simulated cooling technique.

4.6.2.3 Tabu Search

Tabu search was developed by F. Glover in 1986 [26], who said that Tabu search "*leads a local search process to explore beyond local optimum*". It is based upon the mechanism used by human memory. Here lies the main difference between this search and simulated annealing techniques, the acquired knowledge from the past. The basic operating tenets of Tabu search are simple: a unique solution is employed, which is updated through consecutive iterations. The transition from the current solution "*i*" to the new one "*i* + *1*" at each iteration entails two states:

- Firstly, a group of possible solutions is generated {neighbourhood, Neighbourhood (i)}, which is obtainable from the current solution by means of an incremental movement.
- Each of the possible solutions integrated into the neighbourhood {f(Neighbourhood(i))} is evaluated and the solution able to minimise the objective function will be chosen. It is of importance to remark that this selection is done without taking into account if the value reached by the objective function manages to improve the solution of ith state or not. And thus, it can lie beyond local optimum values and strategically searching solutions.

To keep the process from coming back to old local optimum values, Tabu search classifies a determined number of the most recent movements as "Tabu movements", which cannot be repeated within a specific timeframe. This fact allows for the algorithm to run away from local optimum values in a systematic and non-random way. The memory of past events allows Tabu search to alter the search neighborhood of the current solution and hence to modify the searching process.

Memory structures can take several forms: they can store total information (explicit memory) or just partially, keeping information of certain attributes changing from one solution to the other (attributive memory). This allows choosing those events to be memorized, as well as the ones to be forgotten.

Along with the above-mentioned short-term memory, able to store information regarding those events recently arisen, two additional mechanisms are employed: intensification and diversification, which support an algorithm with a long-term memory. These mechanisms use, mainly, the information corresponding to frequency and the time that a specific attribute has stayed at different solutions during the searching process. Intensification implies the search of certain areas of solution space, in which it is assumed that the global optimum could be found. For its part, the diversification promotes the search through certain areas of the solution space, whose attributes have been poorly used till now, with the main objective of leading the searching process through different areas. By assigning weights to the different attributes of best solutions, it is possible to explore areas that are of great interest.

In comparison with simulated annealing, Tabu search uses fewer parameters and thus it becomes easier to implement. However, the employment of a mechanism such as diversification or intensification increases the method's complexity. An implementation's example of this method to reactive power planning can be found in [27].

4.6.2.4 Evolutionary Algorithms

Natural evolution is an optimization process based on populations. Evolutionary algorithms differ from conventional optimization methods by jointly handling objective and constraints. Theoretically, these techniques converge to a global optimum of solutions with probability 1.

Evolutionary algorithms arose in the late 1950s and their application has steadily increased due to an improvement of calculating capability and a decrease in the cost of computer equipment, as well as due to the development of parallel massive architectures on which these processes are based.

The basic tenet of an evolutionary algorithm is quite simple: a number "N" of individuals is randomly selected from the searching area and this group will be considered as the initial population. Further on, each individual will be analysed to determine its adaptation grade to the environment. Following generations will be obtained by implementing mutation, recombination, reproduction, crossover and selection of the working population. Mutation allows randomly modifying an attribute of one individual; recombination combines the information from several individuals; reproduction enables retention of the best individual's attributes through consecutive generations; crossover makes possible random interchange of information from two individuals; and finally, selection allows removing the worst adapted individuals from the population. As it can be observed, each operator highlights one of the evolutionary facets.

Evolutionary algorithms are classified depending on the element upon which genetic operators are applied:

- *Evolutionary programming*: change is done according to population, employing rules of transitory probability to select the offspring in such a way that each offspring competes against individuals belonging to the previous population, as well as against the resulting population from the mutation process. The winners of previous population will take part in the new population.
- *Evolutionary strategies:* change is done according to individuals by implementing mutations upon those parents selected for the reproduction process and introduced into the new population depending on some selection's variables.

- *Genetic Algorithm (GA)*: carries out operations according to chromosomes. GA stands out as a useful tool to solve multi-objective optimisation problems with continuous or discrete variables. The search of the optimum is performed through a population, instead of using just an individual, and this allows quickly exploring the solution space. These algorithms use only information related to objective function without the need of calculating either derivatives or gradients. Finally, GAs put into practice probabilistic rules of transition to guide the searching process.

Evolutionary algorithms are mainly implemented when heuristics cannot provide any solution or if this solution is not satisfactory. Compared to non-lineal programming, evolutionary algorithms are indeed the best methods to solve optimization problems whose optimization function is not continuous and abrupt.

Similar to the metaheuristics already mentioned in previous sections, there are in literature numerous implementation examples of these evolutionary algorithms solving problems regarding reactive power planning. In [28] the implementation of evolutionary strategies to solve the problem of reactive power planning is described, as well as a comparison between evolutionary algorithms and lineal programming. Finally, references [29, 30] contain examples of ways to handle reactive power planning by implementing GAs and giving some sort of improvement to these optimization algorithms.

GAs differ from traditional mathematical methods of optimization by four key points:

- GAs work with codifications of variables belonging to the problem subject of optimization, as well as operate jointly with multiple parameters.
- GAs operate from a population made up of possible solutions, instead of using just a potential solution. In other words, GAs are intrinsically parallel. This characteristic makes GAs an optimization method able to run away from local optimum values, since it allows the algorithm to work with different searching directions, instead of just one, and hence exploring multiple directions of the solution space. Those searching directions that are not productive enough are quickly removed by genetic operators as happens in nature with the worst adapted individuals. This parallelism enables GAs to work within large searching spaces in which the implementation of intensive searching methods would not otherwise bring about any result within a reasonable timeframe.
- GAs do not use either derivatives or other properties of objective function, as traditionally methods do. They only employ the objective function itself. This characteristic allows GAs to work by means of discontinuous, loud, time-dependent functions or with multiple local optimum values. The crossover genetic operator plays a crucial role in the run away from the local optimum, since it makes it possible to transfer information between profitable candidates of consecutive generations.
- GAs observe rules related to probabilistic transition and are not deterministic.

Among evolutionary computer techniques, GAs are the most widely used for the following reasons:

- Basic ideas of evolutionary approach are compiled in natural ways inside this technique by using a main genetic operator: recombination or crossover.
- They are flexible and easy to adapt to a huge amount of different problems belonging to diverse areas. GAs could be combined with other non-evolutionary methods bringing about hybridization between optimization techniques.
- GAs have a larger theoretical basis. They are based on the schema theory of Holland in 1975. A detailed description of this theory can be found at [31].
- They are quite versatile, as to start operating they need less specific knowledge of the problem.
- It is possible to implement them by means of a medium-capacity computer system, and obtain acceptable results.

4.7 Example

Considering the system in Fig. 4.1, it illustrates a four bus micro grid with one load on each bus. Total active and reactive load of the system is 500 MW and 309.86 Mvar respectively. This micro grid has a tie line to the local utility, slack bus, at bus #4 too. The objective to be pursued is to connect a wind farm with a maximum reactive power capability of 250 Mvar in the optimum bus in order to maximize the loadability of the system without exceeding, at the same time, an admissible margin of 5 % around nominal voltage supply (230 kV) as it is stated in the utility regulations.

4.7.1 Initial Population

The first step in the implementation of the GA is to generate the initial population. There are several methods to obtain this population. In this case, and taking into account the limits of the different variables of the system, the randomize option has been selected as the most suitable one. For the example given, population is composed by five individuals with three genes each one, that correspond to the loading parameter, the wind farm bus allocation and the reactive power injection respectively.

Table 4.1 shows the initial population. It could be observed that individual #1 corresponds to the chromosome 1 that will locate the wind farm in bus # 2, with a reactive power injection of 149.537 Mvar and with a loadability factor of 13.2 %. The loadability parameter represents the systems overload, having in mind the initial loading condition.

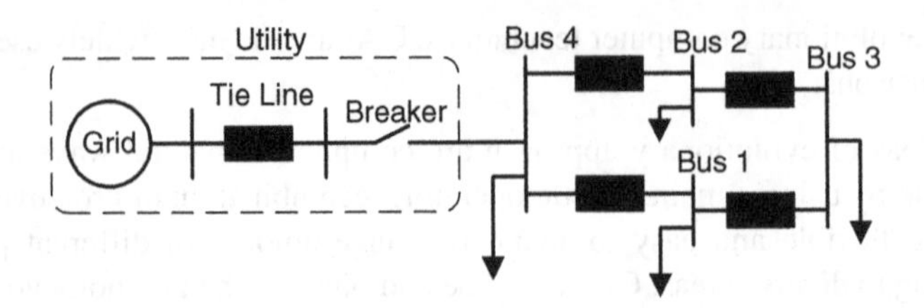

Fig. 4.1 Four bus microgrid

Table 4.1 Initial population of the 4 bus system

	λ(p.u.)	Bus_{WF}	Q_{WF}(Mvar)
Chromosome 1	0.13205	2	149.537
Chromosome 2	0.69965	2	187.455
Chromosome 3	0.4859	2	233.818
Chromosome 4	0.18272	3	209.539
Chromosome 5	0.10121	3	221.811

4.7.2 Evaluation

The evaluation process assigns a fitness value to each individual of the population in terms of the fitness function FF(y) = λ, (see columns two and three of Table 4.2) according to the objective function (4.27). After finishing the evaluation process, a scaling process is applied to the individuals of the population by using the range operation (4.28). The best individual of the population, which has bigger fitness value (FF), will be assigned the top range value of one.

$$\min \quad F(y) = (1 - FF(y)) \tag{4.27}$$

$$range = \frac{1}{\sqrt{i}} \tag{4.28}$$

4.7.3 Selection

Selection operator is applied to the population in order to obtain the group of parents. In this section, the roulette method is used. The first step in the selection process is to determinate the frequency of each individual in terms of their ranges. After that, the cumulative frequency of each individual is calculated (Table 4.3) taking into account that it represents the probability of each individual and could be shown in a circle diagram (Fig. 4.2)

Once the cumulative frequency of each chromosome has been calculated, a random number "r" (between 0 and 1) is associated to each individual of the

Table 4.2 Evaluation process of 4 bus system

	FF(y)	F(y)	Order	Range
Chromosome 1	0	1.0000	3	0.557
Chromosome 2	0	1.0000	4	0.350
Chromosome 3	0	1.0000	5	0.407
Chromosome 4	0.18272	0.8173	1	1.000
Chromosome 5	0.10121	0.8988	2	0.707

Table 4.3 Cumulative frequency of 4 bus system

	Range	Frequency	Cumulative freq.
Chromosome 1	0.557	0.1787	0.1787
Chromosome 2	0.500	0.1547	0.3333
Chromosome 3	0.407	0.1384	0.4718
Chromosome 4	1.000	0.3094	0.7812
Chromosome 5	0.707	0.2188	1.0000

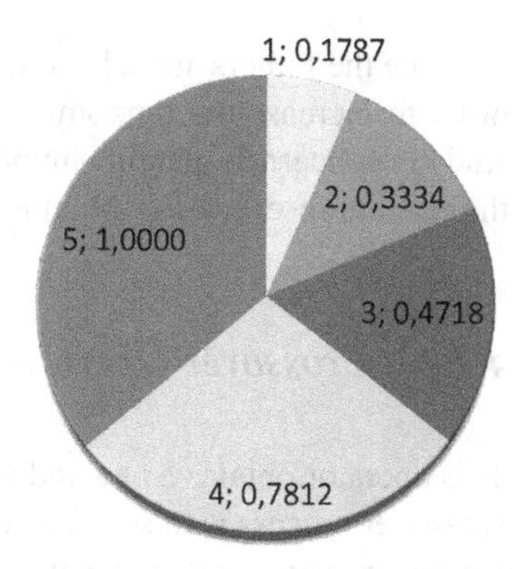

Fig. 4.2 Cumulative frequency diagram

Table 4.4 Selection process of the 4 bus system

	Cumulative freq	*r*	Father
Chromosome 1	0.1787	0.8631	5
Chromosome 2	0.3333	0.3807	3
Chromosome 3	0.4718	0.749	4
Chromosome 4	0.7812	0.1567	1
Chromosome 5	1	0.0581	1

population. In the given example, parents groups are formed by five individuals, so that, five random numbers are chosen (column r of Table 4.4). The selection of each parent depends not only on the random number, r, but also on the cumulative frequency. For each chromosome, its selected parent is the one whose cumulative frequency is immediately above to its random number r. In Table 4.4, it could be seen that for chromosome 1, which has a random number $r = 0.8631$, the parent selected is the one whose cumulative frequency is superior to r, which corresponds to chromosome number 5.

Table 4.5 Parents of the selection process of the 4 bus system

	λ(p.u.)	Bus_{WF}	Q_{WF}(Mvar)	
Chromosome 5	0.10121	3	221.811	Crossover process
Chromosome 4	0.18272	3	209.539	
Chromosome 1	0.13205	2	149.537	
Chromosome 3	0.48590	2	233.818	
Chromosome 1	0.13205	2	149.537	Mutation process

Table 4.6 First couple of parents

First couple of parents			
	λ(p.u.)	Bus_{WF}	Q_{WF}(Mvar)
Chromosome 5	0.10121	3	221.811
Chromosome 4	0.18272	3	209.539

Once the fathers are selected, a random reorganization of the parents is made in order to increase the randomness of the whole process (Table 4.5). As far as this study is concerned, parents involved in the crossover process are the first four, and the rest are reserved for the mutation process.

4.7.4 Crossover Operation

This operator obtains a new individual from a couple of parents. The first step in the crossover process is to select a couple of parents and a random point to perform the crossover operations. In the example, the selected couple of parents are the two first individuals, chromosomes #5 and #4, and the selected crossover point is located between genes 1 and 2; so that, a new offspring individual could be created copying gen 1 from the individual #5 and genes 2 and 3 from the individual #4 (Table 4.6). For the second couple of parents, the random point to perform the crossover operations is located between genes 2 and 3, and thus the new offspring is created by copying genes 1 and 2 from the individual #1 and gen 3 from individual #3 (Table 4.7). The final solution is shown in Table 4.8.

4.7.5 Mutation

Mutation process incorporates new information in the evolution process. To simplify the case study, in the example high mutation rate has been chosen ($p_m = 0.5$). At the beginning of the mutation process a random number "m" between 0 and 1 is assigned to individuals that participate in the mutation process. In the studied case will be chromosome #1. After that, mutation rate is compared with the random number of each individual; if the random number is lower, all genes of the individual will mutate.

Table 4.7 Second couple of parents

Second couple of parents			
	λ(p.u.)	Bus_{WF}	Q_{WF}(Mvar)
Chromosome 1	0.13205	2	149.537
Chromosome 3	0.48590	2	233.818

Table 4.8 Evolution of the crossover process

	Parent			Children			
	λ(p.u.)	Bus_{WF}	Q_{WF}(Mvar)	λ(p.u.)	Bus_{WF}	Q_{WF}(Mvar)	
Chromo. 5	0.1012	3	221.81	0.1012	3	209.53	Child1
Chromo. 4	0.1827	3	209.53				
Chromo. 1	0.1320	2	149.53	0.1320	2	233.81	Child2
Chromo. 3	0.4859	2	233.81				

Table 4.9 Mutation process of the 4 bus system

Parent				Child		
λ(p.u.)	Bus_{WF}	Q_{WF}(Mvar)	m	λ(p.u.)	Bus_{WF}	Q_{WF}(Mvar)
0.13205	2	149.537	0.05	0.89842	3	151.358

Table 4.9 shows the mutation process of the example. Parent population is made of five individuals; four of them are used in the crossover process, and thus, only one parent is used in the mutation process which corresponds to individual #1. The random number m, associated to individual #1, is 0.05, lower than the mutation rate ($p_m = 0.5$), so that, all genes of the individual will mutate and a new offspring will be born.

4.7.6 New Population

After the application of the evaluation, selection, crossover and mutation process, a new population is created. This new population (Table 4.10) is formed by two individuals obtained by the crossover operation, one for the mutation process, and the other two individuals are the result of applying a new operator called elitism. In that case, elitism operator selects the two individuals of the population which have obtained the top fitness value across the evolution process. In the studied example, the two individuals selected by the elitism process are chromosomes #4 and #5 which are included in the new population.

4.7.7 Final Solution

This evolution process is applied till one solution has been found or one of the stop parameters have been reached. The most common considered stop criteria is to

Table 4.10 New population of the 4 bus system

Genetic operator	λ(p.u.)	Bus_{WF}	Q_{WF}(Mvar)
Elitism children	0.18271	3	209.539
	0.10121	3	221.810
Crossover children	0.10121	3	209.539
	0.13205	2	233.817
Mutation child	0.89842	3	151.358

Table 4.11 Evolution of the population of the 4 bus system

		λ(p.u.)	Bus_{WF}	Q_{WF}(Mvar)	F
Original population		0.132	2	149.54	1
		0.699	2	187.46	1
		0.485	2	233.82	1
		0.182	3	209.54	0.8173
		0.101	3	221.81	0.8988
Population 1	Elitism	0.182	3	209.54	0.8173
		0.101	3	221.81	0.8988
	Crossover	0.101	3	209.54	0.8988
		0.132	2	233.82	1
	Mutation	0.898	3	151.36	1
Population 2	Elitism	0.182	3	209.54	0.8173
		0.101	3	221.81	0.8988
	Crossover	0.182	3	209.54	0.8173
		0.101	3	151.36	0.8988
	Mutation	0.132	3	213.05	0.8671
Population 3	Elitism	0.182	3	209.54	0.8173
		0.182	3	209.54	0.8173
	Crossover	0.132	3	213.05	0.8671
		0.182	3	209.54	0.8173
	Mutation	0.829	3	135.23	1
Population 4	Elitism	0.182	3	209.54	0.8173
		0.182	3	209.54	0.8173
	Crossover	0.182	3	209.54	0.8173
		0.182	3	209.54	0.8173
	Mutation	0.793	3	155.33	1
...	...				
Population 10	Elitism	0.204	3	282.95	0.7958
		0.204	3	282.95	0.7958
	Crossover	0.182	3	282.95	0.8173
		0.204	3	209.54	0.8173
	Mutation	0.552	2	180.46	1
Best individual		**0.204**	**3**	**282.95**	**0.7958**

reach the limit of the maximum allowable generations or to reach the convergence tolerance between two consecutive populations. Table 4.11 shows the evolution process of the example.

It could be remarked that the optimal final solution consists in adding one wind farm at bus #3 with a reactive power capability of 282.95 Mvar and thus, increasing the loadability of the system in a 20 %.

References

1. Pudjianto D, Ahmed S, Strbac G (2002) Allocation of var support using LP and NLP based optimal power flows. IEE Proceed Gener Trans Distrib 149(4):377–383
2. Hsu CU, Yan YH, Chen CS, Her SR (1993) Optimal reactive power planning for distribution systems with nonlinear loads. In: TENCON '93. Proceedings. IEEE Region 10 conference on computer, communication, control and power engineering, Beijing, China, vol 5, pp 330–333, Oct 1993
3. Chen Y-L (1996) Weak bus oriented reactive power planning for system security. IEE Proceed Gener Trans Distrib 143(6):541–545
4. Ajjarapu V, Ping Lin Lau S, Battula S (1994) An optimal reactive power planning strategy against voltage collapse. IEEE Trans Power Syst 9(2):906–917
5. Chakrabarti BB, Chattopadhyay D, Krumble C (2001) Voltage stability constrained var planning-a case study for New Zealand. In: 2001 large engineering systems conference on power engineering. LESCOPE '01, Halifax, Canada, pp 86–91
6. Obadina OO, Berg GJ (1989) Var planning for power system security. IEEE Trans Power Syst 4(2):677–686
7. Aoki K, Fan M, Nishikori A (1988) Optimal var planning by approximation method for recursive mixed-integer linear programming. IEEE Trans Power Syst 3(4):1741–1747
8. Delfanti M, Granelli GP, Marannino P, Montagna M (2000) Optimal capacitor placement using deterministic and genetic algorithms. IEEE Trans Power Syst 15(3):1041–1046
9. Degeneff RC, Neugebauer W, Saylor CH, Corey SL (1988) Security constrained optimization: an added dimension in utility systems optimal power flow. Comput Appl Power IEEE 1 (4):26–30
10. Gomez T, Perez-Arriaga IJ, Lumbreras J, Parra VM (1991) A security-constrained decomposition approach to optimal reactive power planning. IEEE Trans Power Syst 6(3):1069–1076
11. Chattopadhyay D, Bhattacharya K, Parikh J (1995) Optimal reactive power planning and its spot-pricing: an integrated approach. IEEE Trans Power Syst 10(4):014–2020
12. Lee KY, Bai X, Park Y-M (1995) Optimization method for reactive power planning by using a modified simple genetic algorithm. IEEE Trans Power Syst 10(4):1843–1850
13. Granville S, Pereira MVP, Monticelli A (1988) An integrated methodology for var sources planning. IEEE Trans Power Syst 3(2):549–557
14. Zhu JZ, Chang CS, Yan W, Xu GY (1998) Reactive power optimisation using an analytic hierarchical process and a nonlinear optimisation neural network approach. IEE Proceed Gener Trans Distrib 145(1):89–97
15. Chen YL, Ke YL (2004) Multi-objective var planning for large-scale power systems using projection-based two-layer simulated annealing algorithms. IEE Proc Gener Trans Distrib 151 (4):555–560
16. Jwo W-S, Liu C-W, Liu C-C, Hsiao Y-T (1995) Hybrid expert system and simulated annealing approach to optimal reactive power planning. IEE Proc Gener Trans Distrib 142(4):381–385
17. Iba K (1993) Reactive power optimization by genetic algorithm. In: Conference proceedings on power industry computer application conference, Scottsdale, AZ, USA, pp 195–201, May 1993
18. Hsiao Y-T, Chiang H-D, Liu C-C, Chen Y-L (1994) A computer package for optimal multi-objective var planning in large scale power systems. IEEE Trans Power Syst 9(2):668–676

19. Mantovani JRS, Garcia AV (1996) A heuristic method for reactive power planning. IEEE Trans Power Syst 11(1):68–74
20. Thomas WR, Dixon AM, Cheng DTY, Dunnett RM, Schaff G, Thorp JD (1995) Optimal reactive planning with security constraints. In: 1995 I.E. conference proceedings on power industry computer application conference, Salt Lake City, UT, USA, pp 79–84, May 1995
21. Venkataramana A, Carr J, Ramshaw RS (1987) Optimal reactive power allocation. IEEE Trans Power Syst 2(1):138–144
22. Kazempour SJ, Conejo AJ (2012) Strategic generation investment under uncertainty via benders decomposition. IEEE Trans Power Syst 27(1):424–432
23. Kirkpatrick S, Gelatt C, Vecchi M (1983) Optimization by simulated annealing. Science 220(4598):498–516
24. Dreo J (2006) Metaheuristics for hard optimization: methods and case studies. Springer, Berlin
25. Hsiao Y-T, Liu C-C, Chiang H-D, Chen Y-L (1993) A new approach for optimal var sources planning in large scale electric power systems. IEEE Trans Power Syst 8(3):988–996
26. Glover F (1986) Future paths for integer programming and links to artificial intelligence. Comp Operat Res 13(5):533–549
27. Zhang W, Liu Y, Liu Y (2002) Optimal var planning in area power system. In: International conference on power system technology. Proceedings. PowerCon 2002, Kunming, China, vol 4, pp 2072–2075
28. Lee KY, Yang FF (1998) Optimal reactive power planning using evolutionary algorithms: a comparative study for evolutionary programming, evolutionary strategy, genetic algorithm, and linear programming. IEEE Trans Power Syst 13(1):101–108
29. Ajjarapu V, Albanna Z (1991) Application of genetic based algorithms to optimal capacitor placement. In: Proceedings of the first international forum on applications of neural networks to power systems, Seattle, WA, USA, pp 251–255, July 1991
30. Dong Z-Y, Hill DJ (2000) Power system reactive scheduling within electricity markets. In: 2000 international conference on advances in power system control, operation and management. APSCOM-00, Hong Kong, China, vol 1, pp 70–75, Oct–1 Nov 2000
31. Goldberg DE (1989) Genetic algorithms in search, optimization, and machine learning. Addison-Wesley, Reading

Chapter 5
Voltage Stability in Power Networks with Wind Power Generation

Beyond any doubt, we may consider century 21st as the one devoted to renewable energy. According to the International Energy Agency (IEA) [1], Renewable Energy Sources shall provide about 35 % of the European Union's (EU) electricity by 2020, and within this context, wind energy is set to contribute the most – nearly 35 % – of all the power coming from renewable sources. This evolution is based on sustainability scenarios, like the BLUE one [2] related to the reduction of greenhouse emissions. However, the appropriate integration of such renewable energy into power networks still presents major challenges to Power Systems Operators (PSO) and planners.

Under heavily stressed situations power network could became unstable and lead to voltage collapse. Voltage stability incidents in power networks with intermittent power generation have been experienced over the world. The majority of the voltage stability analysis or voltage collapse focuses on power systems with deterministic parameters. There are little pieces of work that address the voltage stability analysis of power networks with wind energy generation.

The objectives of this Chapter are twofold; Firstly to analyze the voltage stability problem in power networks which are heavily stressed and secondly, to show that wind energy sources coupled to the network through power converters offer the ability to provide a very fast dynamic Var injection, and thus, their optimal allocation in the power network could alleviate the voltage instability or even prevent voltage collapse.

5.1 Voltage Stability Definition and Concepts

According to IEEE/CIGRE Power System Stability definitions [3], it could be said that Voltage Stability refers to the power system ability to maintain steady-state voltages at all buses of the system after being subjected to a disturbance in a given initial operating condition.

H. Amaris et al., *Reactive Power Management of Power Networks with Wind Generation*, Lecture Notes in Energy 5, DOI 10.1007/978-1-4471-4667-4_5,

The main factor causing voltage instability is the power system inability to satisfy the reactive power load demand keeping, at the same time, the voltage at acceptable levels under stressed situations. Usually, voltage stability acts as a load driven and for that reason is often called load stability [4].

According to IEEE definitions 'voltage collapse is the process by which the sequence of events accompanying voltage instability leads to a or abnormal low voltages in a significant part of the power system' [4]. In some situations, the result of the load tap changers, the load dependence to the voltage or the generator reactive power limitations, among others factors, could also generate very low-voltage values in a large part of the system originating, as a result, a partial or total collapse.

Voltage stability can be analysed by using two different approaches: a dynamic and a static one [5]. Dynamic analysis performs time domain simulations and considers dynamic models of the power system (loads, generators, control actions, dynamic models of compensators) to solve nonlinear differential algebraic equations. This methodology provides accurate time-domain results and it seems to be the best approach to analyse specific voltage collapse situations. However, it presents the disadvantage of being time consuming and computationally expensive and consequently this methodology is not usually applied to determine voltage stability limits.

Static analysis is based on solving algebraic equations such as the conventional power flow equations and it offers the advantage of being computationally very efficient and so that this methodology is normally used to determine which are the weak areas prone to voltage instability. When applying static analysis tools to study a dynamic phenomenon such as voltage stability, it has to be noted that the results are limited to specific time moments considering that the system is subjected to small disturbances. In that case, this approach is able to give information about the stability margin (how close the operation point is to voltage collapse) or to identify the point where voltage instability occurs.

5.2 Voltage Stability in a Two-Bus Power System

Consider a Wind Farm connected to an infinite bus through a lossless transmission line as shown in Fig. 5.1. The wind farm is injecting active (P_{WT}) and reactive power (Q_{WT}) at BUS 2 respectively. A single load, connected to BUS 2, demands active power (P_D) and reactive power (Q_D).

For the sake of simplicity, the load is assumed to behave as an impedance, in which consumed power by the load does not depend on frequency or on voltage variations in BUS 2.

The infinite bus (BUS 1) is represented as an ideal voltage source $\vec{E}$ in which voltage and frequency are constant. We assume three-phase and steady-state sinusoidal operating conditions, consequently, the phasor voltage source is $\vec{E} = E\angle 0$.

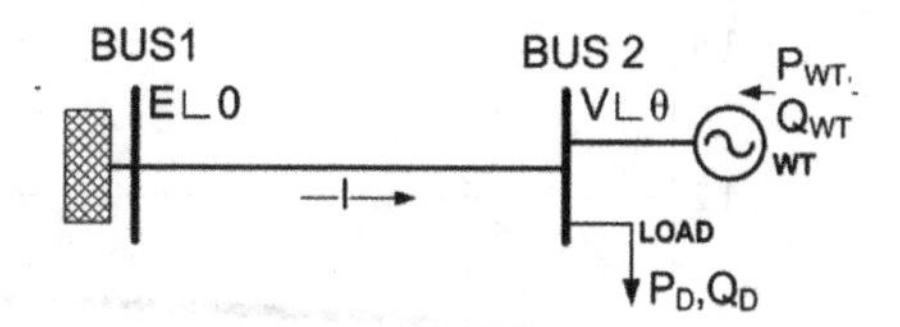

Fig. 5.1 A single-two-bus system with a wind farm

Voltage instability could be produced when loads try to draw more power than the one able to be delivered by the transmission and by the generation system [6]. The more the load increases, the more the voltage in the load bus decreases until reaching a critical value that corresponds to the maximum power transfer. This maximum power transfer is related to voltage instability. Beyond this point voltage, stability is lost and voltage collapse could easily occur.

From Fig. 5.1

$$U = \vec{E} - jX\vec{I} \tag{5.1}$$

The absorbed complex power at bus 2 is:

$$\vec{S} = P + jQ = U\vec{I}^* = U\frac{\vec{E}^* - U^*}{-j} = \frac{j}{X}(EU\ \cos\ \theta + jEU\ \sin\ \theta - U^2) \tag{5.2}$$

where:

$$P = P_D - P_{WT}$$

$$Q = Q_D - Q_{WT}$$

This decomposes into:

$$P = -\frac{EU}{X}\sin\ \theta \tag{5.3}$$

$$Q = -\frac{U^2}{X} + \frac{EU}{X}\cos\ \theta \tag{5.4}$$

Eliminating θ gives:

$$(U^2)^2 + (2QX - E^2)U^2 + X^2(P^2 + Q^2) = 0 \tag{5.5}$$

This is a second-order equation with regard to U^2. The condition to have at least one solution is:

$$(2QX - E^2)^2 - 4X^2(P^2 + Q^2) \geq 0 \tag{5.6}$$

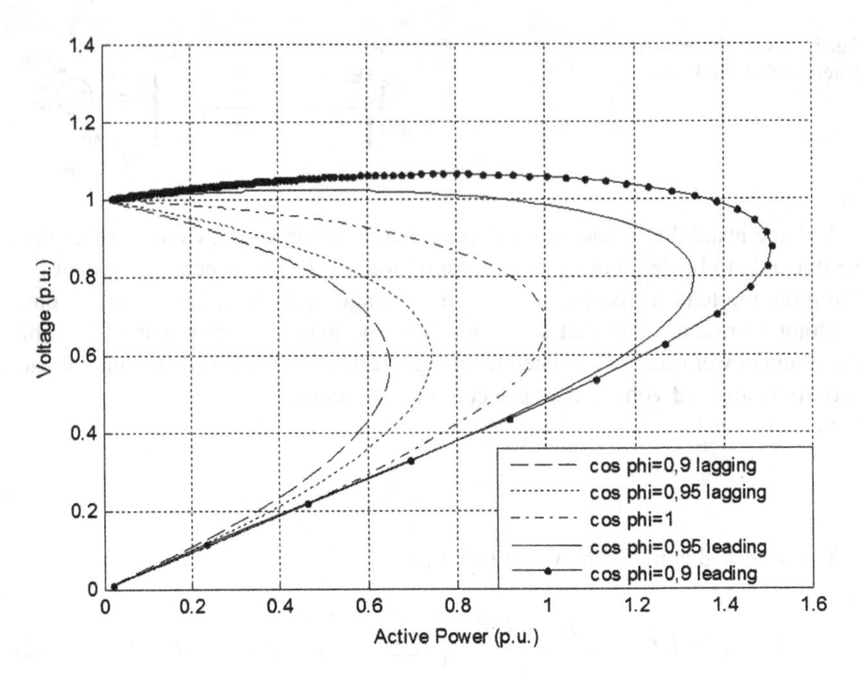

Fig. 5.2 PV curves

The two solutions are given by:

$$U = \sqrt{\frac{E^2}{2} - QX \pm \sqrt{\frac{E^4}{4} - X^2P^2 - XE^2Q}} \tag{5.7}$$

If the wind farm is not operating, the active and reactive power injected by the wind farm are $P_{WT} = Q_{WT} = 0$. In this situation active and reactive power in BUS 2 correspond only to the consumed power load, $P = P_D$ and $Q = Q_D$.

Consumed reactive power by the load depends on the active power and the load power factor by $Q = P \tan \phi$. By means of this expression it is possible to obtain curves of load voltage just as a function of active power (PV) for various $\tan \varphi$ (Fig. 5.2). These curves are known as PV curves and they allow calculating the relationship between voltages and load in a specific region as soon as the load starts increasing.

We can define the quantities and U_{maxP} using the PV diagram (Fig. 5.3). P_{max} is the maximum deliverable power and U_{maxP} is the voltage in which this maximum happens.

This P_{max} is often called the Point of Voltage Collapse (PoC) where the voltage drops rapidly with an increase of load. The risk of voltage instability can be measured by calculating the distance of the initial operating point (base case) to the Point of Voltage Collapse. This distance is called Voltage Stability Margin (VSM).

Fig. 5.3 PV curves

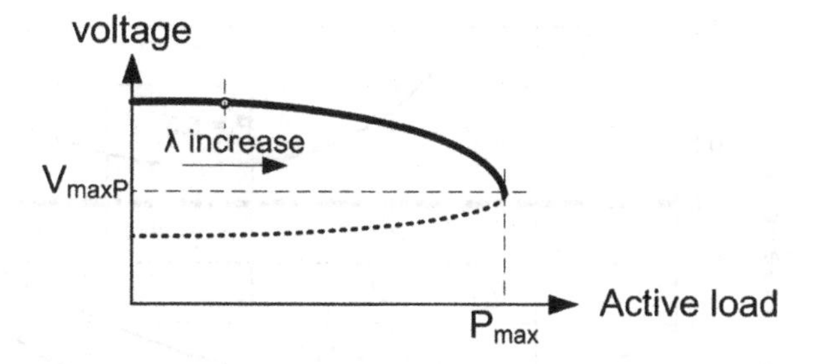

5.3 Voltage Stability Enhancement by Wind Farms

When the connected Wind Farm to BUS 2 has at its disposal reactive power injection facility, Q_{WT}, the maximum deliverable power P_{max} increases, as well as the voltage stability margin could also rise. In this case, the load flow equations become:

$$P = -\frac{EU}{X}\sin\theta \tag{5.8}$$

$$Q - Q_{WT} = -\frac{U^2}{X} + \frac{EU}{X}\cos\theta \tag{5.9}$$

If we assume a constant power load, for each value of the voltage U, θ is first obtained and then the reactive power Q_{WT} is computed. V-Q curves are a set of graphs that describes the relationship of the amount of reactive compensation at a given bus to obtain a desired voltage in that bus.

Three VQ curves are shown in Fig. 5.4 for different load conditions (P = 0.5 p.u; P = 3.0 p.u.; P = 5.3 p.u.). The two intersection points with the V axis correspond to a no-reactive compensation ($Q_{WT} = 0$).

It can be observed that the load situation corresponding to P = 3.0 p.u is a more loaded one than P = 0.5 p.u. In this case, the intersection point **O'**, with no reactive compensation, gives a voltage around 0.95 p.u. at the PCC. In this situation, it will be necessary to inject a certain amount of Q_{WT} (around 0.5 p.u.) in order to restore the nominal voltage.

The third curve represents a load condition (P = 5.3 p.u.) in which the system cannot longer operate without reactive power injection.

The main function of reactive compensation devices is to provide voltage support to avoid voltage instability or a large-scale voltage collapse.

Reactive power capability in fixed speed wind turbines can be implemented in different ways: by connecting shunt capacitors, SVC (static var compensator) and (static synchronous compensator) explained in Chap. 2. Variable speed wind turbines offer voltage control capability at the Point of Common Coupling (PCC) by making use of their reactive power injection capability as was shown in Chap. 2. Incorporating the STATCOM functionality in their control, variable speed wind turbines can be considered as controllable reactive power sources similar to STATCOM.

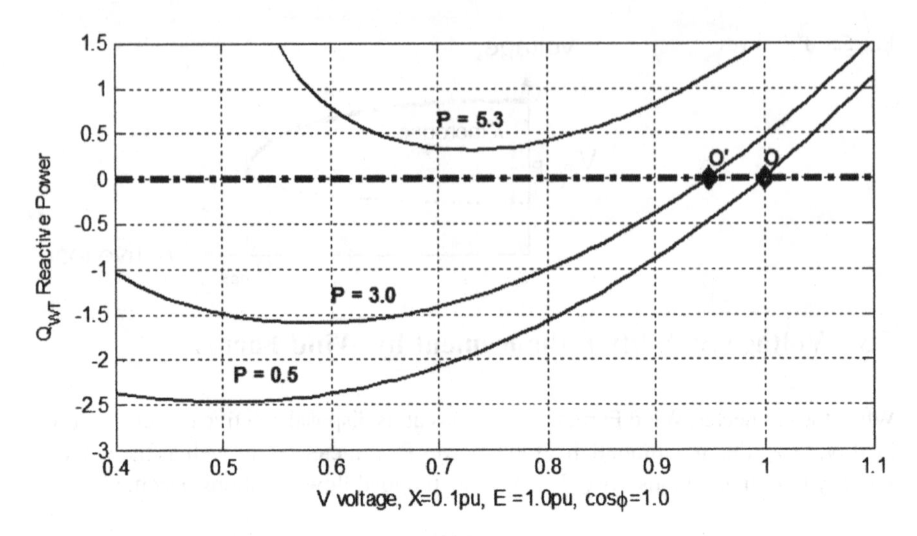

Fig. 5.4 Relationship between voltage and reactive power at BUS 2

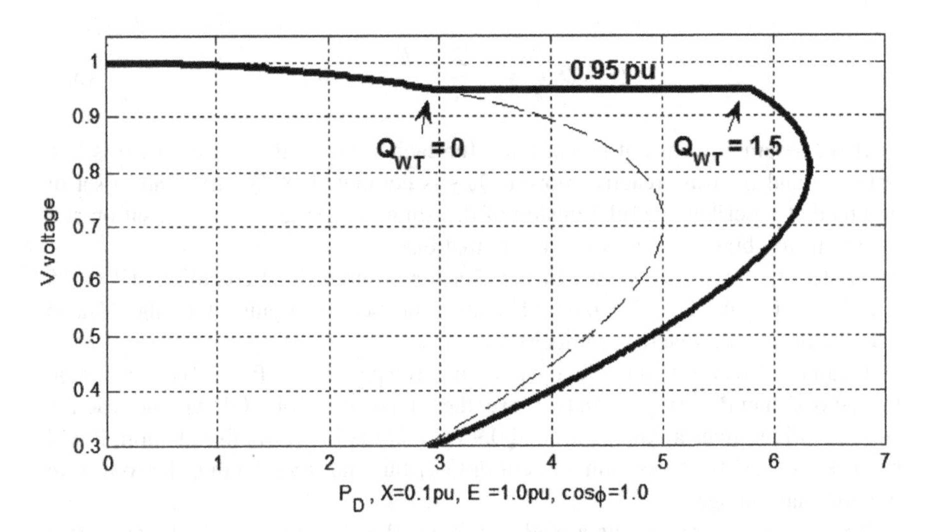

Fig. 5.5 PV curves in a wind farm working as a STATCOM

In Fig. 5.5. PV curves in a wind farm working as a STATCOM, it is shown the given relationship between voltage and the consumed load power (PD) in BUS 2 when the wind farm is operating and offering reactive injection capability as a STATCOM.

The initial situation corresponds to the point in which $Q_{WT} = 0$ (no reactive power injection). In this case, in order to keep the voltage within the wind farm at the value of 0.95 p.u., the maximum load to be connected is $P_D = 3.0$ p.u. It can be observed that the voltage stability margin has increased from the initial situation to maximum injection ($Q_{WT} = 1.5$ p.u.) while maintaining 0.95 p.u. voltage constant.

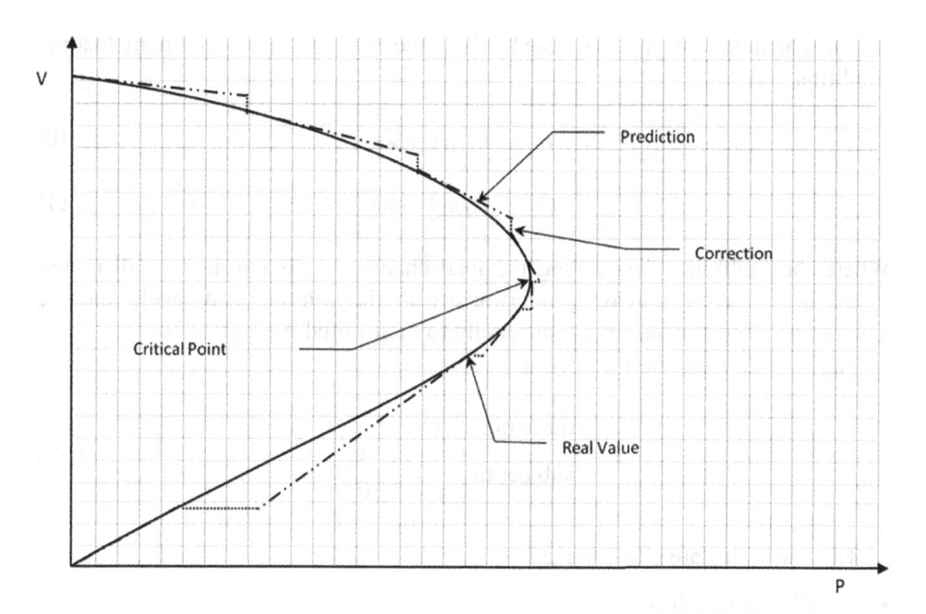

Fig. 5.6 Continuation method

It can be noted that at higher reactive power injection level, the equilibrium operating point progressively approaches the nose point of the PV curve.

5.4 Optimal Voltage Stability

We have already defined loadability limit as the point in which demand reaches the maximum value. If the load is to be considered as constant power, then the loadability limit corresponds to the maximum power deliverable in buses. Voltage stability limits are difficult to obtain by means of the classical load flow methodologies since load flow calculations fail to converge close to the point of voltage collapse (PoC). The classical way to obtain PV and QV curves is by the continuation method [7] which consists of two steps (Fig. 5.6):

- The predictor step: From an initial operating point, find the next point for a load P increase by estimating the changes in the power flow variables.
- The corrector step: to solve the power flow equations for the next operating point by using the estimated values obtained at the predictor step as initial values.

Another way of computing the maximum loadability point corresponds to the use of optimization methodologies [8–10] as the OPF or the heuristic techniques explained in Chap. 4. In these situations, the optimization problem is implemented to represent properly the power system security to the maximum distance to the voltage collapse.

The parameter to be maximized drives the system to its maximum loading condition:

$$P_D = P_{D0}(1+\lambda) \tag{5.10}$$

$$Q_D = Q_{D0}(1+\lambda) \tag{5.11}$$

Where the multiplier designating the load change rate is constant in all nodes. Loads in all buses increase proportionally to their initial load levels and the generators outputs increase proportionally to their initial generations too.

The objective function is:

$$\begin{aligned} &\text{maximize}\, f(x) = \lambda \\ &\text{subject to} \begin{cases} f(x) = 0 \\ g(x) \leq 0 \end{cases} \end{aligned} \tag{5.12}$$

The constrains are classified as:

- Equality Constraints
  Basic equality constrains correspond to the power flow equations in every buses:

$$P_{Gi} - P_{Di} = P_{Lik} \tag{5.13}$$

$$Q_{Gi} - Q_{Di} + Q_{FACTS,i} = Q_{Lik} \tag{5.14}$$

Bus ith is linked to bus kth through a transmission line whose admittance is:

$$Y_{ik} = G_{ik} + B_{ik} \tag{5.15}$$

Voltage buses at buses ith and kth are: $\vec{U}_i = U_i \angle \theta_i$ and $\vec{U}_k = U_k \angle \theta_k$
where $\theta_{ik} = \theta_i - \theta_k$
Active and reactive transmission power between buses ith and kth:

$$P_{Lik} = U_i \sum_{i=1}^{N} U_k (G_{ik} \cos\theta_{ik} + B_{ik} \sin\theta_{ik}) \tag{5.16}$$

$$Q_{Lik} = U_i \sum_{i=1}^{N} U_k (G_{ik} \sin\theta_{ik} - B_{ik} \cos\theta_{ik}) \tag{5.17}$$

- Inequality Constraints
  - Network technical constraints: Voltage level at buses is not allowed to be outside the maximum and minimum values according to grid voltage regulations.

$$U_{i,min} \leq U_i \leq U_{i,max} \qquad i = 1, 2, \ldots, N_B \tag{5.18}$$

  - Physical limits of generators: The maximum injected power by wind farms is constrained by the maximum capability of each wind turbine. Active power injected by the wind turbine will depend on the wind availability for each wind turbine.
    Active power output is restricted by lower and upper limits: where $P_{gi,min}$, $P_{gi,max}$ are the minimum and maximum operating power respectively:

$$P_{gi,min} \leq P_{gi} \leq P_{gi,max} \qquad i = 1, 2, \ldots, N_G \tag{5.19}$$

$$Q_{gi,min} \leq Q_{gi} \leq Q_{gi,max} \qquad i = 1, 2, \ldots, N_G \tag{5.20}$$

  - Physical limits of FACTS devices: Reactive power from FACTS devices is restricted by the lower and upper limits:

$$Q_{FACTSi,min} \leq Q_{FACTSi} \leq Q_{FACTSi,max} \quad i = 1, 2, \ldots, N_{FACTS} \tag{5.21}$$

  - Wind farm connection point (LocWF) constraints: The potential connection point from the wind farm to the grid is limited to the geographical area in which the available wind resource is higher:

$$Bus_{LocWFi,min} \leq Bus_{LocWFi} \leq Bus_{LocWFi,max} \quad i = 1, 2, \ldots, N_G \tag{5.22}$$

5.5 Case Studies

5.5.1 IEEE-14 Bus Power System

The OPF formulation is applied to a modified IEEE-14 bus power system [11] in which the original synchronous generator in bus 8 has been removed and a wind farm has been installed with a rated power $P_{WT} = 30$ MW (Fig. 5.7).

Two different scenarios have been considered:

- Scenario 1: The wind farm in bus 8 is adjusted to maintain a constant power factor, $\cos\varphi = 1$ ($Q_{WT} = 0$).
- Scenario 2: The wind farm in bus 8 offers reactive power capability similar to a SVC or device. It can also be considered as a controllable reactive power source.

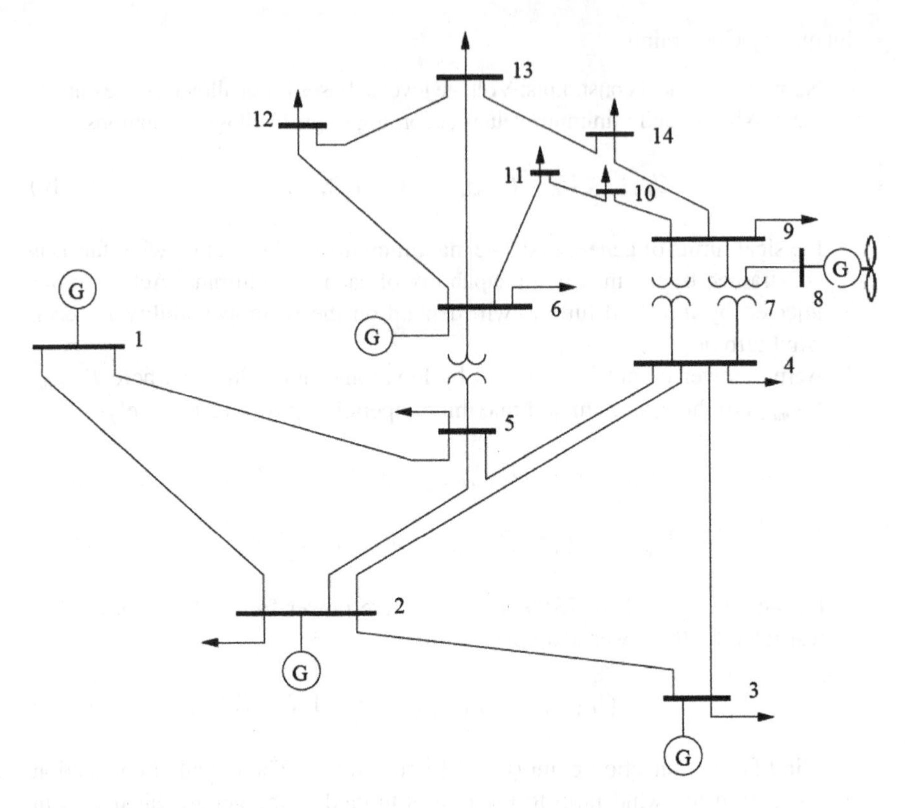

Fig. 5.7 Modified IEEE 14-bus test system

Figure 5.8 shows the voltage profile at IEEE-14 buses considering the initial condition, which corresponds to a wind farm connected in bus 8 with unity power factor ($Q_{WT} = 0$). It can be observed that voltage levels in all buses are within $\pm$ 1 0% of nominal voltage. Bus #3 is the bus with lower voltage levels.

Optimal Power Flow formulation is applied to both scenarios and results are shown in Tables 5.1 and 5.2.

Figure 5.9 shows the voltage evolution versus λ load parameter for scenario 1, critical buses are #3, #10, #13, and #14. It can be seen that the maximum loadability of the system if keeping voltages at all buses within $\pm$ 10% of nominal voltage, corresponds to $\lambda = \lambda = \lambda_{OPFMAX} = 0.45$. This result implies a maximum loadability of 374 (MW). Moreover, it is to be noted that bus #14 is the most critical one when reaching the established limits in the OPF formulation.

The maximum loadability in scenario 2 (Fig. 5.10) increases compared to scenario 1. It can be also seen that the load parameter, $\lambda = \lambda = \lambda_{OPFMAX} = 0.66$ implies a maximum loadability of 430 (MW). This second scenario involves the reactive injection capability of the wind farm. It can be noted how the reactive power injection at the wind farm terminals improves the system voltage stability.

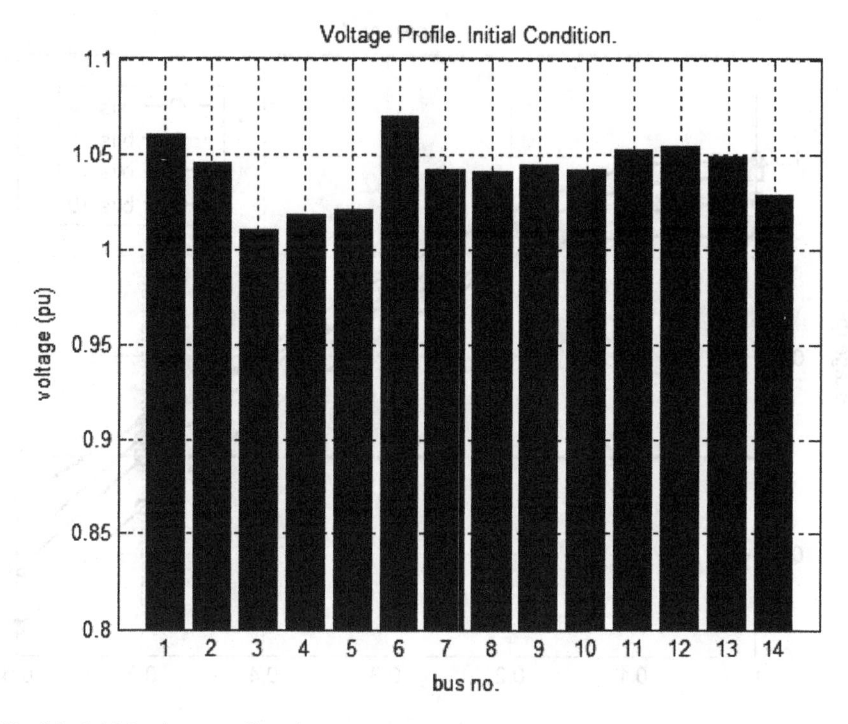

Fig. 5.8 Initial voltage profile at buses

Table 5.1 Optimal loadability in scenario 1

Scenario 1	$\lambda\ \lambda_{\mathrm{OPF}_{\mathrm{MAX}}} = \lambda_{critical}$	Maximum loadability (MW)	Critical bus #
Initial condition	0	259	–
OPF solution	0.45	374	14

Table 5.2 Optimal loadability in scenario 2

Scenario 2	$\lambda\ \lambda_{\mathrm{OPF}_{\mathrm{MAX}}} = \lambda_{critical}$	Maximum loadability (MW)	Critical bus #
Initial condition	0	259	–
OPF solution	0.66	430	14

This numbers should be interpreted as a higher capability of the power system to handle disturbances; consequently, the voltage stability improves.

As a conclusion, it can be noted until what extent wind energy penetration can be increased by incorporating reactive power capability at wind farms and by using optimisation algorithms in order to determine the maximum amount of power to be integrated into the system and, at the same time, keeping steady-state voltage profile within a fixed predetermined range under normal and post contingency situations.

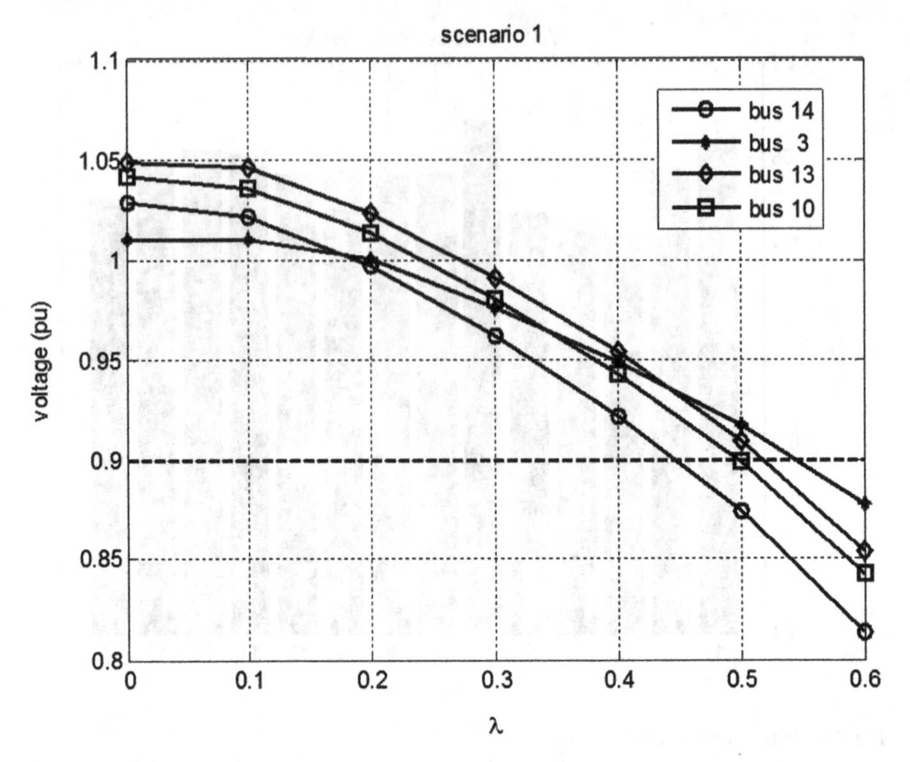

Fig. 5.9 Voltage versus lambda (scenario 1)

5.5.1.1 Load Modelling Influence

It is well known that load modelling has a critical effect on voltage stability studies [12, 13]. The first important aspect to consider is the selection of a static or a dynamic models. A static load model is to be applied is those situations where load power demand does not change significantly on time, and therefore, it is able to describe the relationship between active and reactive power at any time with the voltage and/or frequency at the same instant of time.

On the contrary, a dynamic load model has to consider the relationship of both active and reactive power at any instant of time, as a function of the voltage and/or frequency time history.

Common static load models express active and reactive power demand as an exponential relationship regarding the voltage:

$$P_d = P_{d0}\left(\frac{U}{U_0}\right)^{a_1} \tag{5.23}$$

$$Q_d = Q_{d0}\left(\frac{U}{U_0}\right)^{a_2} \tag{5.24}$$

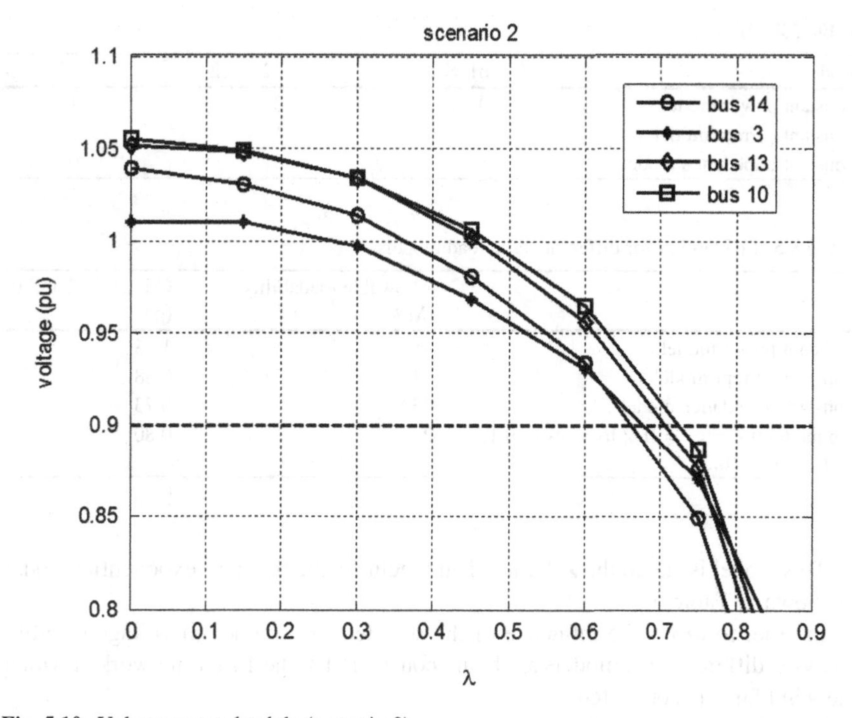

Fig. 5.10 Voltage versus lambda (scenario 2)

Table 5.3 Parameters for static exponential load model

Model	$a_1 = a_2$
Constant power model	0
Constant current model	1
Constant impedance model	2

Where parameters' a_1, a_2 can have different values (Table 5.3):

And U_0 is the nominal voltage, and P_{D0},Q_{D0} are the nominal active and reactive power respectively consumed by the load at the nominal voltage.

A more general static load model can be expressed as a polynomial function such as:

$$P_d = P_{d0}\left[b_1\left(\frac{U}{U_0}\right)^2 + b_2\left(\frac{U}{U_0}\right)^1 + b_3\right] \tag{5.25}$$

$$Q_d = Q_{d0}\left[c_1\left(\frac{U}{U_0}\right)^2 + c_2\left(\frac{U}{U_0}\right)^1 + c_3\right] \tag{5.26}$$

$$b_1 + b_2 + b_3 = c_1 + c_2 + c_3 = 1$$

Table 5.4 ZIP coefficients

Model	b1 = c1	b2 = c2	b3 = c3
Constant power model	0	0	1
Constant current model	0	1	0
Constant impedance model	1	0	0

Table 5.5 Loadability for different static load models

	Allowable loadability [MW]	Critical loadability (p.u.) λ_{lim}
Constant power model	31.75	1.13
Constant current model	28	0.88
Constant impedance model	25.8	0.73
Zip model ($b_1 = c_1 = 0.6$; $b_2 = c_2 = 0.1$; $b_3 = c_3 = 0.3$)	27	0.80

This model is called the ZIP model and includes the previous exponential model as shown at Table 5.4.

In order to prove the influence of the selected load model on voltage stability analysis, different load models are being considered at the 14-bus network at which the wind farm is connected.

It can be noted (Table 5.5) that the constant power model is the one which allows greater loadability margin. On the contrary, if using a constant impedance model the allowable and critical loadability is to be reached with smaller load increases. This result empathizes the importance of selecting a proper load model for voltage stability analysis.

5.5.2 *Application Case*

In this case, the proposed methodology is applied to a distribution network (Fig. 5.11) composed by 34 buses (details are included in the appendix). The aim of the optimization methodology is to optimally locate several wind farms that offer reactive power capability spread over the distribution network.

The chromosome length depends on the number of wind farms, n, to be located and it will have (1 + 2*n) genes.

5.5.2.1 Maximizing the Voltage Stability

In order to maximize the power system loadability, the objective function is defined by (5.12), and the constraints of the problem are the ones defined in (5.13)–(5.22).

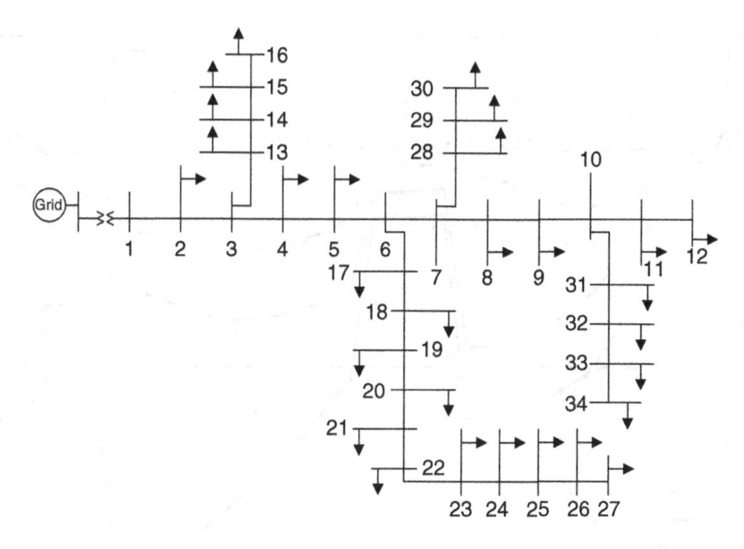

Fig. 5.11 34 -Bus distribution network

Table 5.6 Solution of the GA

Case	$\lambda_{lim.}$(p.u.)	$Bus_{WF\ (\#)}$	Q_{WF} (Mvar)
Base case Without WF	0	–	–
One WF	0.06	27	2.77
Two WF	0.3	11	2.95
		25	2.79
Three WF	0.74	10	2.15
		23	2.45
		26	2.71

Table 5.6 shows the solution proposed by the GA. It could be pointed out that the voltage stability improves as the number of wind farms connected to the grid increases too. In the case of adding three wind farms, the voltage stability increases up to 74 %. Moreover, it could be observed, that the optimal bus location for the connection of the wind farm is quite close in the network, in spite of the fact that it depends on the number on units connected to the system.

Figure 5.12 shows the voltage profile for the different simulations. It could be highlighted that the incorporation of several wind farms with reactive power capability improves the voltage profile of the power systems by smoothing it.

Finally, Fig. 5.13 shows the voltage stability P-V curves of the different cases. It could be remarked that as the penetration level of wind power penetration increases the loadability and the distance to the point of collapse of the system increases too. So, as a conclusion, it could be stated that the incorporation of wind farms in power systems improves voltage stability.

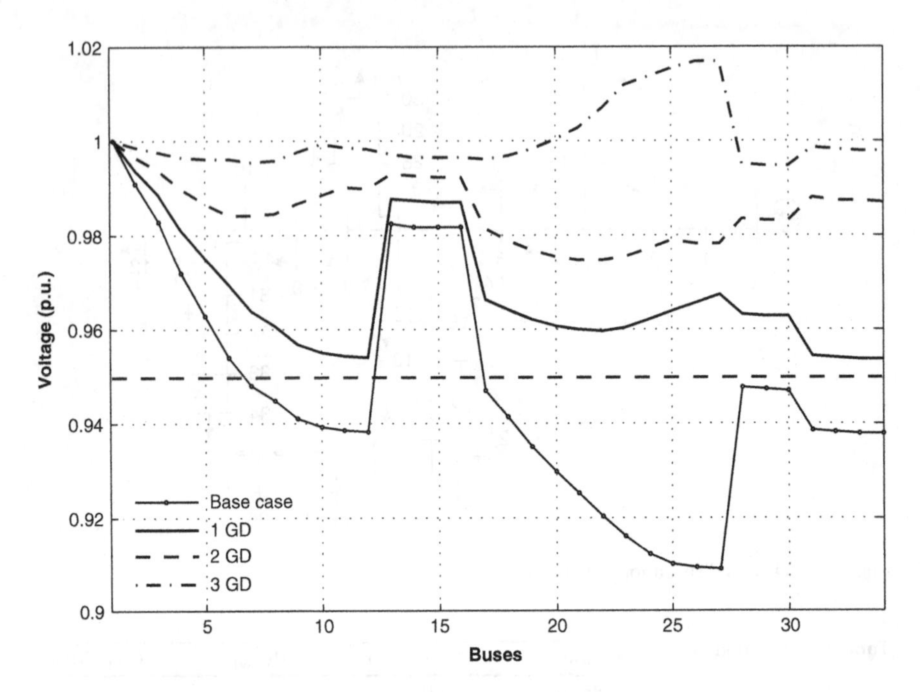

Fig. 5.12 Voltage profile of the 34-bus network

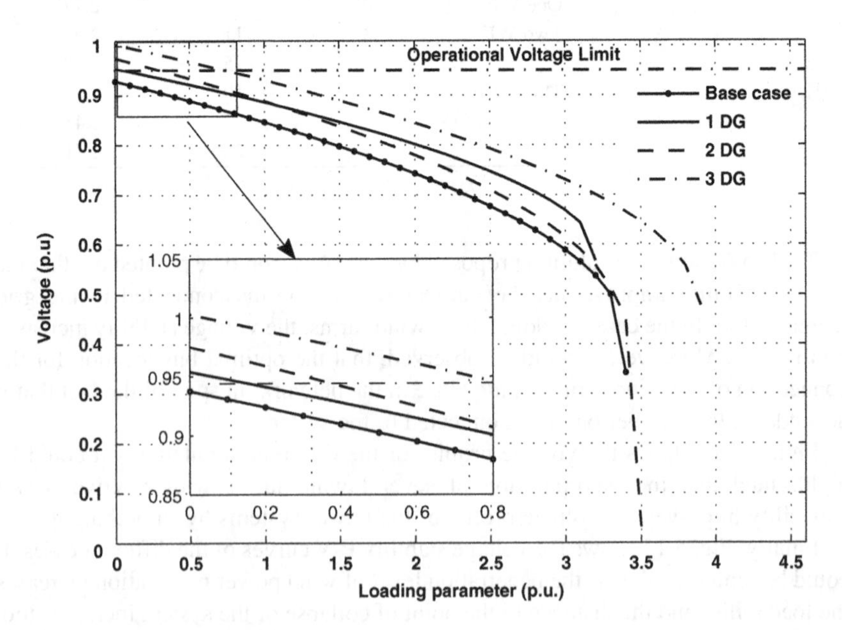

Fig. 5.13 P-V Curves of the 34 bus system

Table 5.7 Maximum loadability and penetration level

λ_{lim} (p.u.)	% Wind power penetration
0.62235	70.09

Table 5.8 Solution of the GA

Loc_1 (#)	P_1 (MW)	Q_1 (Mvar)	Loc_2 (#)	P_2 (MW)	Q_2 (Mvar)	Loc_3 (#)	P_3 (MW)	Q_3 (Mvar)	Loc_4 (#)	P_4 (MW)	Q_4 (Mvar)
10	1.49	1.8	25	1.5	1.72	21	1.44	1.94	22	1.42	2

5.5.2.2 A Multiobjective Approach: To Increase Voltage Stability by Maximizing the Wind Farm Penetration Level

In this study, the main objective of the proposed methodology applied to the 34 buses network is to determine the optimal allocation of four wind farms and, at the same time, to calculate their active and reactive power injection in order to maximize the voltage loadability of the system as well as to increase the wind farm penetration level. In this case, a multiobjective function that incorporates both single objectives is used as follows:

$$\min F(y) = \frac{1}{2}(1 - f(y)) + \frac{1}{2}\left(\frac{1}{g(y)}\right) \tag{5.27}$$

where:

$f(y)$, optimize loading parameter
$g(y)$, represents penetration level

Tables 5.7 and 5.8 show the results of the GA. It could be observed that the maximum power system loadability corresponds to an overload of 62.2 % with a maximum penetration level of 70.09 %. This condition is reached when one wind farm is connected at bus number #10 with an active and reactive power injection of 1.49 MW and 1.8 Mvar respectively (Table 5.8); the second wind farm is connected at bus # 25 with an active and reactive power injection of 1.5 MW and 1.72 Mvar; the third wind farm is connected at bus number #21 and injects 1.44 MW and 1.94 Mvar of active and reactive power respectively, and the last wind farm is connected at bus #22 and injects 1.42 MW of active power and 2 Mvar of reactive power.

Figure 5.14 represents the voltage profile of the power systems in the original situation (base case, without wind farm), and in the case of incorporating four wind farms at their optimal allocation without increasing the load (dash-dot line) as well as in the case of the maximum loadability (solid line, overload of 62.2 %).

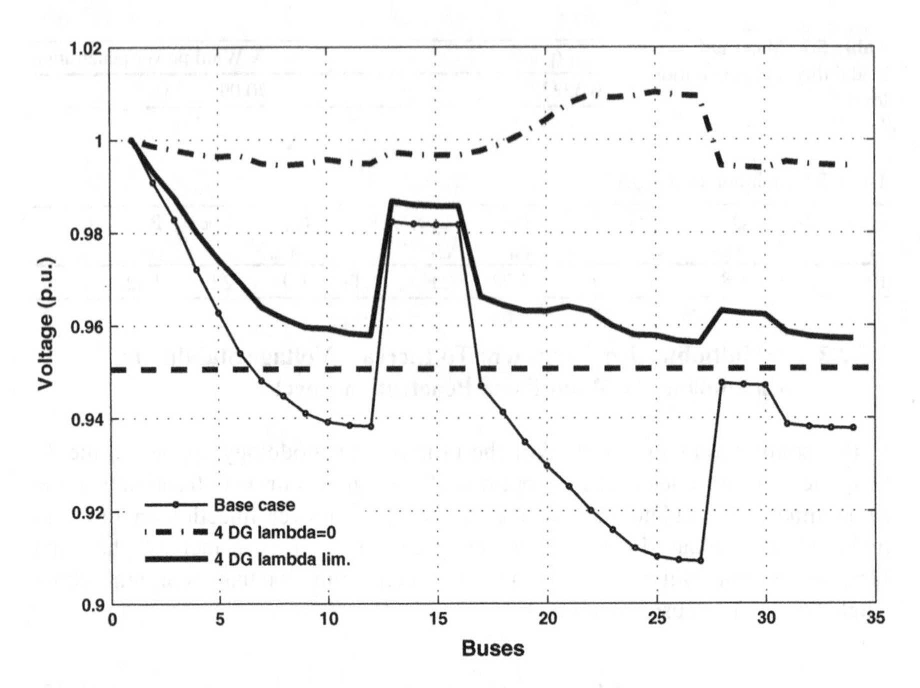

Fig. 5.14 Voltage profile of the 34 bus system for the multiobjective study

References

1. Organization for Economic Cooperation & Devel (2010) Energy technology perspectives: scenarios and strategies to 2050: 2010. Organization for economic cooperation & Devel, Paris, France, July 2010
2. International Energy Agency (2009) Ensuring green growth in a time of economic crisis – the role of energy technology. International Energy Agency, Siracusa, Italy
3. Kundur P, Paserba J, Ajjarapu V, Andersson G, Bose A, Canizares C, Hatziargyriou N, Hill D, Stankovic A, Taylor C, Van Cutsem T, Vittal V (2004) Definition and classification of power system stability IEEE/CIGRE joint task force on stability terms and definitions. IEEE Trans Power Syst 19(3):1387–1401
4. Ajjarapu V (2006) Computational techniques for voltage stability assessment and control. Springer, United State of America
5. Kundur P (1994) Power system stability and control. McGraw-Hill Professional, New York
6. Cutsem TV, Vournas C (1998) Voltage stability of electric power systems. Kluwer, Boston. ISBN 0792381394
7. Canizares CA, Alvarado FL (1993) Point of collapse and continuation methods for large ac/dc systems. IEEE Trans Power Syst 8(1):1–8
8. Qiao J, Min Y, Lu Z (2006) Optimal reactive power flow in wind generation integrated power system. In: Proceedings of the international conference power system technology PowerCon 2006, Chongqing, China, pp 1–5
9. Zhang W, Li F, Tolbert LM (2007) Review of reactive power planning: objectives, constraints, and algorithms. IEEE Trans Power Syst 22(4):2177–2186
10. Mariotto L, Pinheiro H, Cardoso G, Muraro MR (2007) Determination of the static voltage stability region of distribution systems with the presence of wind power generation. In:

Proceedings of the international conference clean electrical power ICCEP '07, Capri, Italy, pp 556–562

11. Abu-Hashim R, Burch R, Chang G, Grady M, Gunther E, Halpin M, Harziadonin C, Liu Y, Marz M, Ortmeyer T, Rajagopalan V, Ranade S, Ribeiro P, Sim T, Xu W (1999) Test systems for harmonics modeling and simulation. IEEE Transactions on Power Delivery. 14(2):579–587
12. Mansour Y (ed) (1990) Voltage stability of power systems: concepts, analytical tools, and industry experience. IEEE Press, New York
13. Carson WT (1994) Power system voltage stability. McGraw-Hill, New York

Chapter 6
Reactive Power Management

Reactive Power Management is a critical issue when dealing with the planning and operation of power networks with high wind energy penetration. Reactive Power Management entails the requested operation and planning actions to be implemented in order to improve the voltage profile and the voltage stability in power networks [1]. An efficient reactive power planning could be obtained by choosing an optimum location of var sources during the planning stage, whereas efficient reactive power dispatch could be achieved by scheduling an optimum regulation of the voltage set point at the generators connection point and at the VAR settings during the reactive power dispatch [2] and [3].

6.1 Reactive Power Planning

Reactive Power Planning (RPP) could be formulated as an optimization process [4] in which the objective function to minimize is:

$$\begin{array}{ll} \min\limits_{x \in R^n} & f(x,u) \\ \text{subject to} & \begin{cases} g(x,u) = 0 \\ h(x,u) \leq 0 \end{cases} \end{array} \tag{6.1}$$

Where:

u are the control variables.

$x \in R^n$stands for all the system steady state variables.

$f(x,u)$ objective function to minimize.

$g(x,u)$ involves the equality constraints.

$h(x,u)$corresponds to the inequality constraints.

H. Amaris et al., *Reactive Power Management of Power Networks with Wind Generation*, Lecture Notes in Energy 5, DOI 10.1007/978-1-4471-4667-4_6,

6.1.1 *Objective Function*

In this section the target of the optimization algorithm is to allocate a specified number of wind farms and SVC's units in a power network in order to improve the voltage stability and the voltage profile of the power network. To overcome any eventual inefficient performing, the algorithm manages to find the optimum location and the optimum reactive power injection among wind farms and SVC's in a coordinated way.

In most of the cases the driving force of voltage instability is the load and that is why the algorithm tries to maximize the loadalibility factor taking into account the minimum voltage allowable limit (normally $U_{min} = \pm 10\% U_N$) according to the utilities grid voltage regulations.

In this situation the load change scenarios, P_D and Q_D could be modified as:

$$P_D = P_{D0}(1 + \lambda) \tag{6.2}$$

$$Q_D = Q_{D0}(1 + \lambda) \tag{6.3}$$

Where:

P_{D0}, Q_{D0} correspond to the original load (base case).
λ is the loadalibility parameter.

6.1.1.1 Constraints

- *Equality Constraints.*
  Basic equality constraints correspond to the power flow equations in every buses:

$$\Delta P_i = Pg_i - Pd_i - P_i \tag{6.4}$$

$$\Delta Q_i = Qg_i - Qd_i - Q_i \tag{6.5}$$

where Pg_i and Qg_i are real and reactive powers of generator at bus i^{th}, respectively; Pd_i and Qd_i the real and reactive load powers, respectively; P_i and Q_i the power injections at the node i^{th}:

- Active and reactive transmission power between buses i^{th} and k^{th}
  Bus i^{th} is linked to bus k^{th} through a transmission line whose admittance is $Y_{ik} = G_{ik} + jB_{ik}$. Voltage buses at buses i^{th} and k^{th} are: $\underline{U_i} = U_i \angle \theta_i$ and $\underline{U_k} = U_k \angle \theta_k$

where:

$$\theta_{ik} = \theta_i - \theta_k \quad (6.6)$$

active and reactive power transmitted through the line between buses i^{th} and k^{th} are:

$$P_{Lik} = U_i \sum_{k=1}^{N} U_k (G_{ik} cos\theta_{ik} + B_{ik} sin\theta_{ik}) \quad (6.7)$$

$$Q_{Lik} = U_i \sum_{k=1}^{N} U_k (G_{ik} sin\theta_{ik} - B_{ik} cos\theta_{ik}) \quad (6.8)$$

- Power consumed by loads connected at bus i^{th}
  In this chapter loads are characterized by a static exponential model, considering a power constant model ($\alpha = \beta = 0$), in which active and reactive power consumed by the load is expressed as:

$$P_{Di} = P_{D0} \left(\frac{U}{U_0} \right)^{\alpha} \quad (6.9)$$

$$Q_{Di} = Q_{D0} \left(\frac{U}{U_0} \right)^{\beta} \quad (6.10)$$

  Where U_0, P_{D0} and Q_{D0} are the reference voltage, active and reactive power at bus i^{th}.
- Reactive power injected by SVC units connected at bus i^{th}
  Reactive power injected by the SVC is:

$$Q_{svc}(\alpha_{svc}) = \frac{U^2}{X_c} - U^2 B_{svc}(\alpha_{svc}) \quad (6.11)$$

- Steady state operation point of variable speed wind farms
  The wind power production for each operating point is considered to be known as is detailed in Chap. 3. The steady state working operation point of the DFIG or full power generators is included in the optimisation algorithm.

– *Inequality Constraints.*
  Inequality constraints constitute the physical limits of the components or operational constraints in the system.

- Voltage limits at buses
  Voltage level at buses are not allowed to be outside the maximum and minimum values according to grid voltage regulations.

$$U_{i,min} \leq U_i \leq U_{i,max} \tag{6.12}$$

- Limits of the loadalibility factor
  Lambda is the loading factor by which the load is increased at all buses and it is considered to be $\lambda \geq 0$.
- Limits on variable speed wind generators
  The maximum power injected by wind farms is constrained by the maximum capability of each wind turbine.
  Active power injected by the wind turbine will depend on the wind availability for each wind turbine. Active power output is restricted by lower and upper limits:

$$P_{gi,min} \leq P_{gi} \leq P_{gi,max} \tag{6.13}$$

  where $P_{gi,min}$, $P_{gi,max}$ are the minimum and maximum operating power.
  The capability constraints are related to the Reactive Power capability for the full power converter or DFIG for each working condition as is detailed in Chap. 3.
- Limits on SVC units
  SVC reactive power generation limit is constrained by the physical range of the firing angle. It is considered that the angle is allowed to be set between a maximum and a minimum value.

$$\alpha_{SVC_{i,min}} \leq \alpha_{SVC_i} \leq \alpha_{SVC_{i,max}} \tag{6.14}$$

- Physical constraints in the wind farm connection point
  The potential connection point from the wind farm to the grid is limited to the geographical area in which the available wind resource is higher.

$$Bus_{LocWF_i,min} \leq Bus_{LocWF_i} \leq Bus_{LocWF_i,max} \tag{6.15}$$

6.1.2 Results for an Existing 140-Bus System

6.1.2.1 Test System Description and Problem Formulation

In the next section the proposed RPP methodology is applied upon a real power system. The system is made up of 140 nodes with five voltage levels from 380, 132, 45, 15 kV to low voltage at 380 V. The structure of the 45 kV network is a ring as it

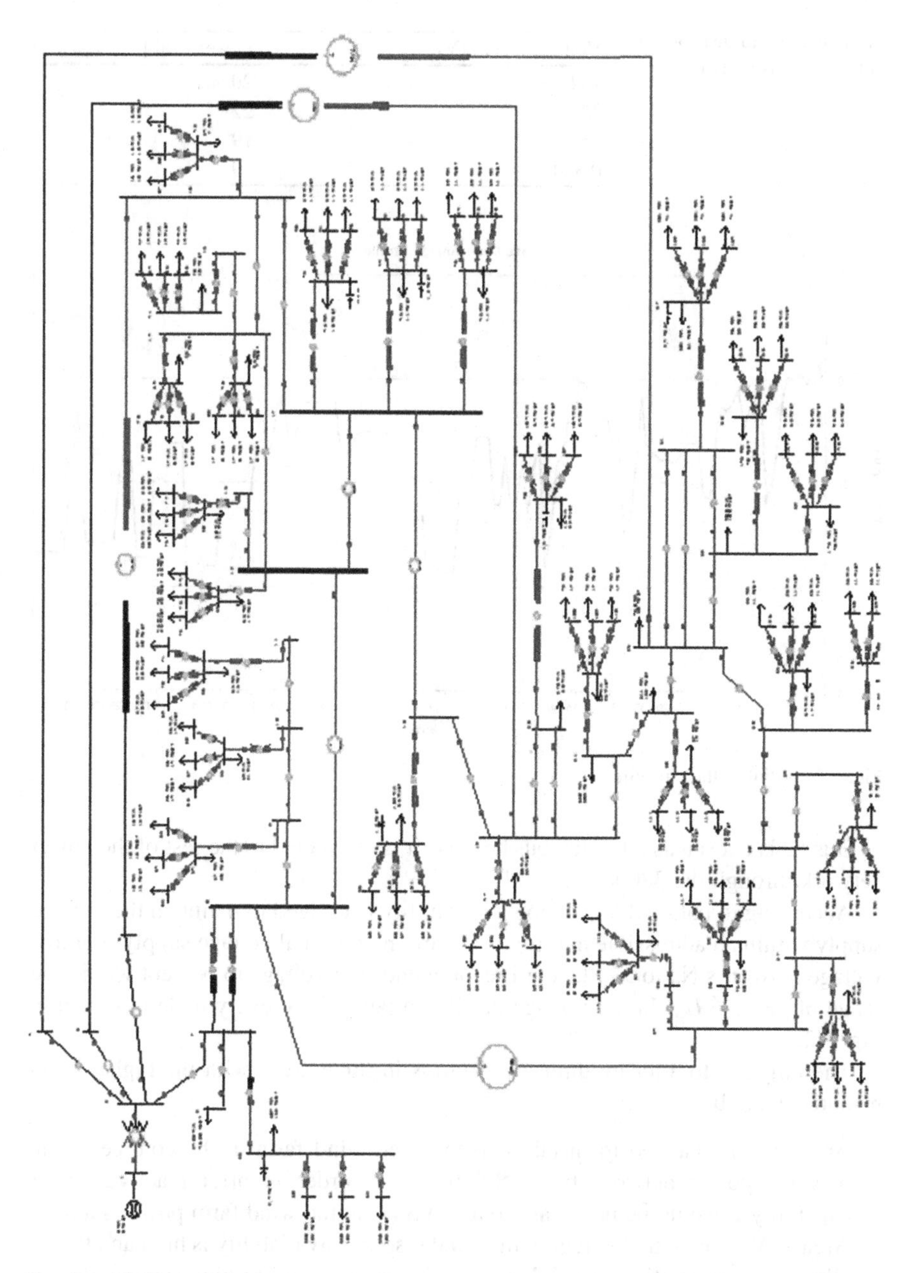

Fig. 6.1 140-Bus test system

can be seen in Fig. 6.1. Although most loads are connected at low voltage level, there are indeed some located at medium voltage level and even a few of them at high voltage level as is shown in Table 6.1.

Table 6.1 Load distribution in the 140-bus system

Voltage level [kV]	Power load [% $P_{total\ load}$]
132	20
45	22
15	19
0.380	9

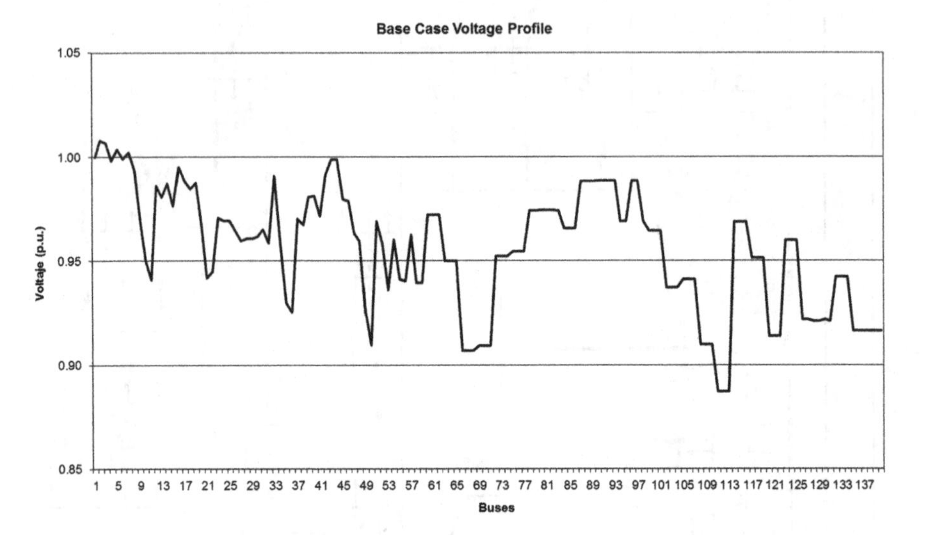

Fig. 6.2 Initial voltage profile

Bus #2 is chosen as the slack bus because it is connected to the rest of the power network through the 380 kV network.

According to national legislation every utility is obliged to maintain the voltage supply within an admissible margin of 7% around nominal voltage supply. For low voltage networks National Standard recommends the voltage to be kept within the range of $\pm 10\%\ U_n$. Initial voltage profile, in per unit, at every node is shown in Fig. 6.2.

The target is to allocate three wind farms in three areas with the higher wind resource availability.

- Area 1: It is planned to install a fixed speed wind farm in the confined zone between power network buses 8-9-10-11. In order to offer reactive power capability it would be necessary to add a SVC at the wind farm power station.
- Area 2 (VSWF2): In this region the wind resource availability is high and thus it allows the installation of a 5 MW variable speed wind turbine (Direct Driven Wind Generator or DFIG). The potential connection points to the network are limited to buses 23-24-25-26-27-28-29-30-31 or 32.
- Area 3 (VSWF1): This area has a very high wind availability resource and therefore it is planned to install a 5 MW variable speed wind turbine. The possible connection points to the power network are restricted to bus numbers12-13-14-15-16-17-18-19-21-21- 22.

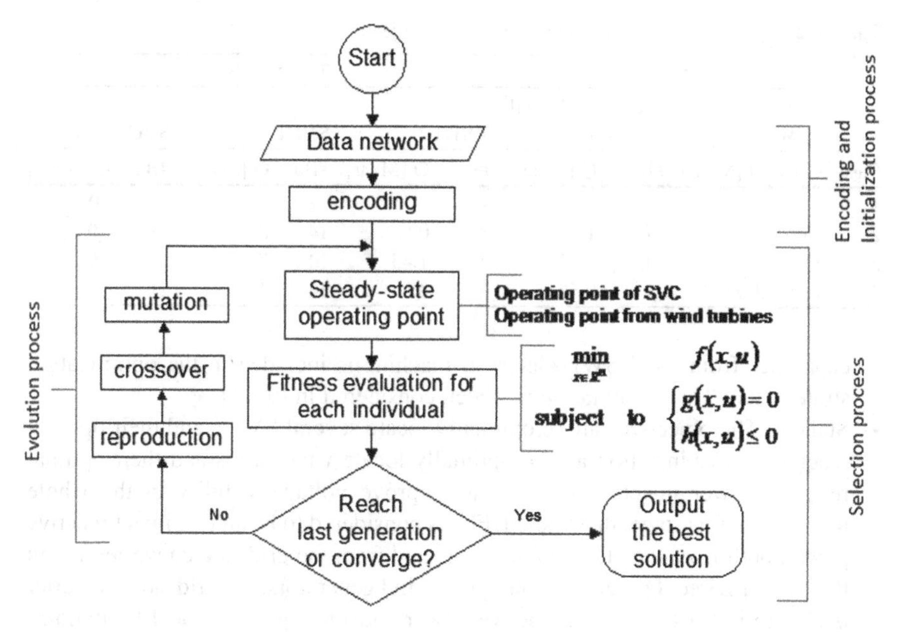

Fig. 6.3 Optimization process

In order to improve the voltage profile, two additional SVC's are to be installed in the whole network. Both, location points and reactive power injection (from the three wind farms and from the two SVC's) will be optimally obtained by the RPP optimization strategy which has been largely explained in previous sections and shown in Fig. 6.3.

The validation approach of the optimization formulation is carried out applying the GA optimization algorithm to different problems increasing gradually the complexity in each stage:

- Stage 1: The objective is to optimally locate several SVC's and to find out the optimal reactive power injection from the SVC's in order to improve voltage stability in the whole network. In this case, there is no wind farm connected to the grid. This is the classical way to perform a RPP study as done in [5] and will be the first problem to be analysed.
- Stage 2: The objectives are to optimally locate several SVC's and their optimal reactive power injection and to optimally locate wind farms in order to improve voltage stability in the whole network. In this stage, VSWF's are modelled as PQ node with a fix power factor that implies that the reactive power injection from the VSWF is fixed and consequently it is not a variable in the RPP algorithm. This approach is the one followed by [6–8].
- Stage 3: The objectives are to optimally locate several SVC's and their optimal reactive power injection and to optimally locate wind farms and their optimal reactive power injection in order to improve voltage stability in the whole network. In this stage, the electrical machine is considered to be able to inject reactive power to the grid and consequently the reactive power injection

Table 6.2 Optimal bus location and reactive power injection

	Three wind farms						Two SVC's			
	VSWF1		Fixed speed wind farm		VSWF2		SVC1		SVC2	
Case	#Bus	Q [Mvar]	#Bus	Q [Mvar]	#Bus	Q [Mvar]	#Bus	Q [Mvar]	#Bus	Q [Mvar]
1	–	–	–	–	–	–	2	8	21	10
2	13	0	11	1	31	0	14	8	21	9
3	13	1.45	11	0.53	31	1.45	14	7	21	9
4	13	3.06	11	0.66	31	3.06	14	4	21	9

capability from the VSWF electrical machine is included in the optimization strategy too. This formulation has been considered in [9–11].

- Stage 4: The objectives are to optimally locate several SVC's and their optimal reactive power injection and to optimally locate wind farms and their optimal reactive power injection in order to improve voltage stability in the whole network. In this stage, the whole DFIG is considered to be able to inject reactive power, both from the electrical machine and from the grid side converter and in the case of Direct Driven variable speed wind generators, the grid side converter is able to inject the needed reactive power considering its physical limitations. The reactive power injection capability from the DFIG (electrical machine + GSC) or full power wind generators is included in the optimization strategy. There are not related works following the same formulation.

6.1.2.2 Simulations

Table 6.2 shows the optimum results obtained by the Genetic Algorithm and the proposed RPP methodology. It should be remarked that in stages 2, 3 and 4 the optimum location of SVC's are always the same. Just in the first case (without wind farms) it will be necessary to change the location of one SVC (from bus #14 to bus #2) in order to improve the voltage profile and voltage stability of the whole network. However, in stage 1, the two SVC's are the only responsible units for injecting reactive power as well as the only ones in charge of improving the voltage stability of the whole network.

VSWF1 are optimally located in bus #13, and VSWF2 are located in bus #31. The optimum location points are not influenced by the reactive power regulation capability available. Fixed speed wind farm is optimally located at bus #11 in all cases.

Furthermore, it should be pointed out that as soon as the reactive power capability of the VSWF starts increasing (stages 2, 3 and 4), the external SVC's located at buses #14, #21 need to inject fewer reactive power. Consequently, the SVC's sizes and cost are reduced.

As can be noted in stages 3–4, the optimization algorithm tries to maximize the reactive power injection available from the VSWF. For that reason, the reactive power injected by the wind farms meets the maximum available value in all cases.

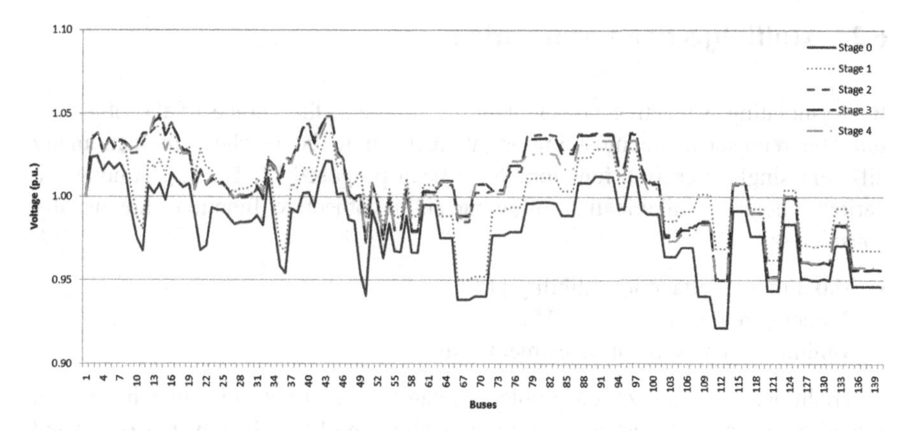

Fig. 6.4 Optimum voltage profile

Table 6.3 Optimal and critical loadability parameter

Stage	$\lambda_{limit}[p.u.]$	$\lambda_{critical}[p.u.]$
Stage 1	0.04	1.2
Stage 2	0.17	1.2
Stage 3	0.18	1.4
Stage 4	0.20	1.4

6.1.2.3 Analysis of Results

Figure 6.4 shows the voltage profile curve for the 140-buses in all the analyzed stages. It should be noted that an increase of the reactive power capability from wind turbines (stages 3 and 4) allows to obtain a voltage profile closer to the voltage reference level (1 [p.u]) with lower voltage dispersion in the whole network.

Table 6.3 shows the voltage stability results: λ_{limit} where the minimum voltage allowable is reached U_{min}; and the $\lambda_{critical}$ where the point of voltage collapse is reached. Results have proved that an increase of the reactive power capability from variable speed wind turbines increases the loadalability parameter and the Voltage Stability Margin too. As a consequence of this widespread increase the whole network improves the voltage stability.

It can be seen that incorporation of reactive power capability of the VSWF (stage 3) improve Voltage Stability Margin (VSM) from 1.2 to 1.4 p.u., so that critical loading corresponding to the point of voltage collapse is increased in a 116.7%. Furthermore, if the reactive power capability of the grid side converter is considered in the optimization process (stage 4), maximum loading of the system, under planning operator voltage limits, is improved in a 111.1%.

6.2 Multiobjective Formulation

By formulating a reactive power planning problem, the choice of the objective function represents the most important decision to be reached. In fact, many different single objective functions have been proposed [2, 12, 13]. Among the various choices available the following single objective functions are usually considered:

- Maximizing voltage loadalibility [14]
- Reducing real power losses [15]
- Minimizing costs in var investment [16]

Traditionally, optimization problems related to multiple objectives have been solved by means of linear programming, where one of the objectives is optimized and the others are included in the restrictions. This procedure generates disadvantages such as:

- Representation of the objectives by means of the restrictions in linear programming can lead to unfeasible problems.
- If the optimization is applied in a large system, it is difficult to find the restriction that does the unfeasibility.
- There is not a clear criterion for choosing the suitable objective function and in many cases the fulfilment of one single objective can be in conflict with others.

Multiobjectives algorithms stand out as a procedure to solve these problems where the optimal solution is obtained by a set of efficient solutions.

$$\min F(y) = \alpha f(y) + \beta g(y) + \gamma h(y) \quad (6.16)$$

Where:

y variable of decision,

$f(y), g(y), h(y)$set of the objectives to satisfy,

α maximize voltage stability,

β minimize active power losses,

γ minimize cost of var investment.

Using different weights α, β y γ as is shown in Table 6.4 allows to obtain several solutions.

In Table 6.5 it can be noted that if the single objective function G.1 is considered as the only goal to fulfil the GA will locate the variable speed wind farm VSWF 1 (area 1) in bus number #10 consuming reactive power; VSWF 2 is located at bus # 32 and the reactive power injected is 3.08 Mvar. The fixed speed wind farm is located at bus #21 and the SVC connected at the substation would have a rating of 8 Mvar. The two additional SVC spread throughout the network are connected at bus #24 and 32 with a total SVC capacity of 17 Mvar. Similar results are obtained by using the single objective function G.2 (reducing active power losses) or G.3

Table 6.4 Weights of multiobjective goals

	α	β	γ
G.1	1	0	0
G.2	0	1	0
G.3	0	0	1
G.4a	0	1/3	2/3
G.4b	1/2	1/4	1/4
G.4c	2/5	1/5	2/5
G.4d	1/3	1/3	1/3
G.4e	1/4	1/4	1/2

Table 6.5 Results of the GA for each goal

	VSWF1		VSWF2		Fixed speed WF		SVC1		SVC2	
	PC1	Q1 (Mvar)	PC2	Q2 (Mvar)	PC3	Q3 (Mvar)	PC4	Q4 (Mvar)	PC5	Q5 (Mvar)
G. 1	10	−0.75	32	3.08	21	8	24	7	32	10
G. 2	9	0.249	32	1.642	21	7	19	7	57	5
G. 3	9	0	29	2	14	2	22	8	41	1
G. 4a	9	2	30	3	21	4	22	1	23	3
G. 4b	9	2	31	2	22	6	49	3	102	1
G. 4c	9	2	32	3	22	3	21	2	120	1
G. 4d	9	0	32	2	22	4	24	1	32	5
G. 4e	9	0	31	3	21	7	33	1	31	2

(reducing var costs). In each situation, the GA will find out the optimal location for the wind farm, the SVCs and for the injected reactive power.

It is to be pointed out that wind farm location are very similar (VSWF1) or are very close (VSWF2, fixed speed wind farm) whatever the strategy function is considered. The main differences starts appearing just with the optimum bus location of the SVC's and the optimum rating.

In the case of the multiobjetive approach, it could be noted that the necessary rating of the SVC are greatly decreased, falling from (7 + 10) Mvar achieved by the application of the single objective function G.1 to (1 + 2) Mvar when the multiobjective function G.4c is applied.

Different aspects are shown in Table 6.6 depending on: each goal function, the voltage stability margin, the active power losses and the cost of the connection of the var sources (represented by the rating of the SVC units).

Figure 6.5 shows the voltage profile for the optimization configuration corresponding to individual goals and the better multiobjective goal, G.4c.

6.3 Reactive Power Ancillary Service

The best trade-off solution found out during the optimization process corresponds to the multiobjective approach, through which it is possible to obtain simultaneously a good voltage stability margin, a minimum var cost and a reduction of

Table 6.6 Maximum loadability, active power losses and capacity of the SVC units for different goals

	$\lambda_{crit.}$	Losses	$\Sigma\, Q_{SVC}$
	(p.u.)	(MW)	(Mvar)
G.1	1.25	3.39	25
G.2	1.2	2.38	19
G.3	1.1	2.87	11
G.4a	1.1	2.53	8
G.4b	1.1	2.63	10
G.4c	1.2	2.5	6
G.4d	1.1	2.61	10
G.4e	1.2	2.52	10

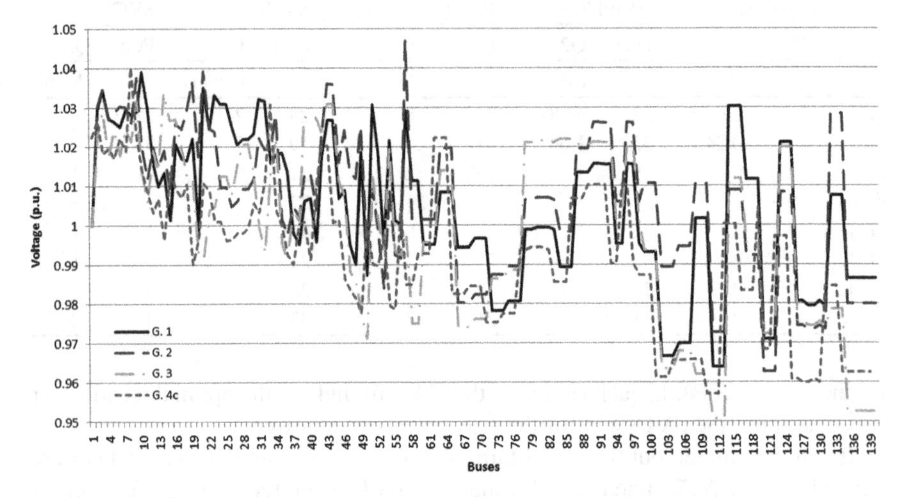

Fig. 6.5 Voltage profile of 140 bus system for different goals

active power losses. Among all the different multiobjective options, the goal function G.4c has been selected as the best trade-off solution.

Power network has to be designed considering the objective functions and the constraints previously shown in section 6.2 but at the same time, it is crucial to take into account the variability from the load power demand between day (peak) and night (valley) and the differences of wind power production between minimum (no production) to the maximum (100% Pn). The analysed scenarios are show in Table 6.7.

For each scenery the multiobjective G.4c approach finds out the optimal location and optimal reactive power injection for each wind farm and SVC's unit (Table 6.8) by means of:

- Maximizing voltage stability,
- Reducing active power losses,
- Minimizing the cost of shunt var sources,
- Keeping voltage profile within $\pm\, 5\% U_N$ for all the scenarios.

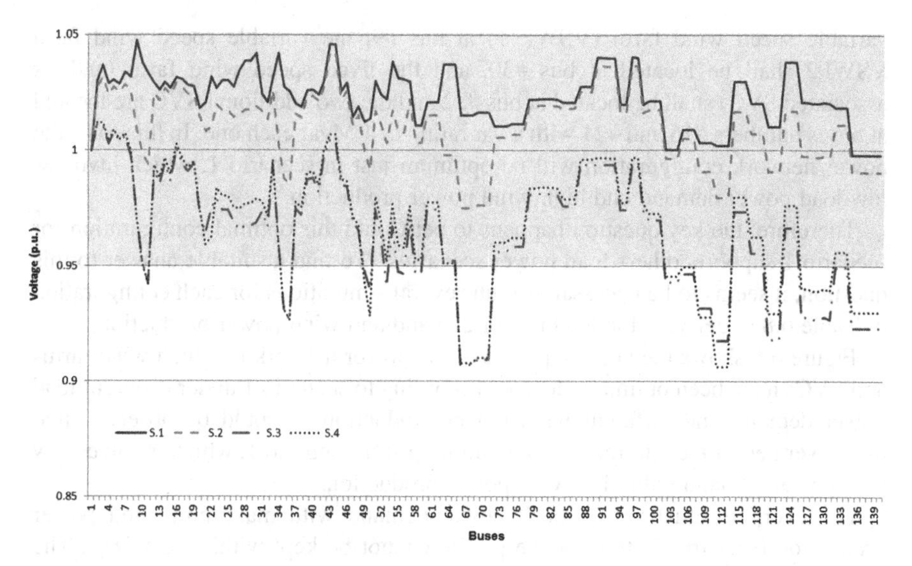

Fig. 6.6 Voltage profile considering the optimal configuration for scenario 1 under load power demand variable and under wind power production variable

Table 6.7 Load power demand – wind power productions scenarios

Scenery	Load power demand	Wind power production
S. 1	Low (valley)	Maximum
S. 2		Minimum
S. 3	High (peak)	Maximum
S. 4		Minimum

Table 6.8 Results of the GA for load power demand – wind power production scenerios

	PC1	Q1 (Mvar)	PC2	Q2 (Mvar)	PC3	Q3 (Mvar)	PC4	Q4 (Mvar)	PC5	Q5 (Mvar)
S. 1	9	−0.82	30	3	22	1	15	1	21	1
S. 2	9	0	25	0	21	1	58	2	10	1
S. 3	9	2	32	3	22	4	58	1	30	1
S. 4	9	0	32	0	22	5	34	3	28	5

6.3.1 Valley Power Demand and Maximum Wind Power Production

If the Transmission System Operator (TSO) tries to design a power network considering one of the scenarios, e.g.: scenario 1 (low load power demand and maximum wind power production) the multiobjetive approach G.4c will locate the variable speed wind farm (VSWF 1) at bus #9, the variable speed wind farm VSWF2 shall be located at bus #30, and the fixed speed wind farm (and its associated SVC) shall be located at bus #22 and the two additional SVC are located at buses numbers #15 and #21 with a var rating of 1 Mvar each one. In this case, the power network configuration will be optimum just in scenario 1, which involves low load power demand and high wind power production.

Therefore, the key question happens to be: Could this optimal configuration for scenario 1 supports others load power scenarios? To find a suitable answer to this question, it seems to be necessary to run several simulations for each configuration of Table 6.8 under variable load power demand and wind power production.

Figure 6.6 shows the voltage profile of the power network in which wind farms and SVCs have been optimally located according to scenario 1 under different load power demand and different wind power production. It could be observed that the power network configuration is optimum just in scenario 1, which involves low load power demand with high wind power production.

However, in situations of peak power demand with maximum wind power production (scenario 3) the voltage profile cannot be kept within $\pm 5\%U_N$. The same behaviour is to be noted under high power demand with minimum wind power production (scenario 4).

6.3.2 Valley Power Demand and Minimum Wind Power Production

If the optimal configuration during the planning stage is obtained considering only the information of scenario 2 (low load power demand with minimum wind power production), the multiobjetive approach G.4c will locate the variable speed wind farm (VSWF 1) at bus #9, the variable speed wind farm VSWF2 shall be located at bus #25, the fixed speed wind farm (and its associated SVC) shall be located at bus #21 and the two additional SVCs are to be located at buses number #10 with a var rating of 1 Mvar and at bus number #58 with a var rating of 2Mvar respectively as deduced from Table 6.8.

Figure 6.7 shows the network voltage profile under different load power demand and wind power production. It is to be observed that once again this configuration can neither keep the voltage level within $\pm 5\%U_N$ under situations of peak power demand with maximum wind power production (scenario 3) nor under peak power demand with minimum wind power production (scenario 4).

6.3.3 Peak Power Demand and Maximum Wind Power Production

If the optimal configuration is obtained considering only the information of the scenario 3 (peak power demand with maximum wind power production), the

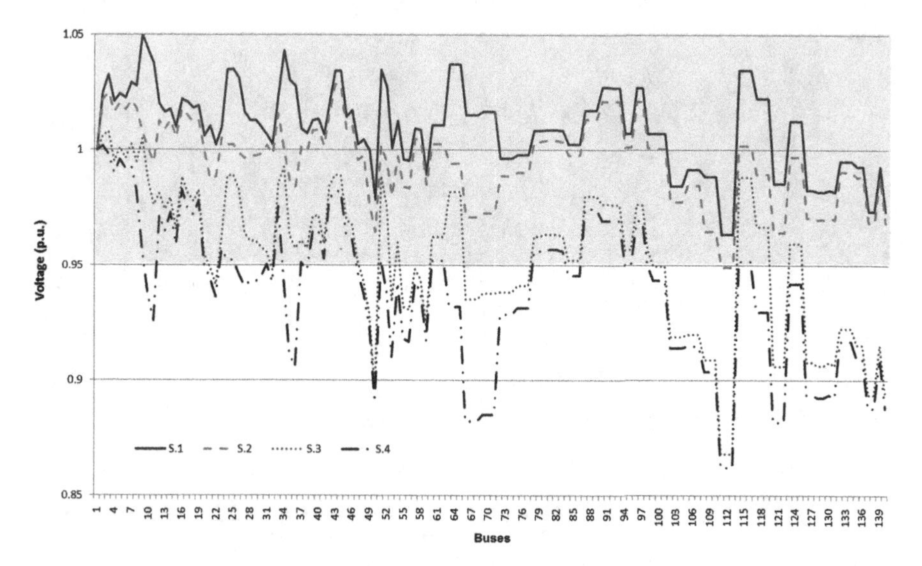

Fig. 6.7 Voltage profile considering the optimal configuration for scenario 2 under load power demand variable and under wind power production variable

multiobjetive approach G.4c will locate the variable speed wind farm (VSWF 1) at bus #9, the variable speed wind farm VSWF2 shall be located at bus #32, the fixed speed wind farm (and its associated SVC) shall be located at bus #22 and the two additional SVCs are to be located at buses number #30 and #58 respectively with a var rating of 1Mvar each one as deduced from Table 6.8.

Figure 6.8 shows the network voltage profile under different load power demand and wind power production. It could be noted that this configuration is not able to keep the voltage level within $\pm\, 5\% U_N$ under situations of peak power demand with minimum wind power production (scenario 4).

6.3.4 Peak Power Demand and Minimum Wind Power Production

If the optimal configuration is obtained considering only the information of the scenario 4 (peak power demand with minimum wind power production), the multiobjetive approach G.4c will locate the variable speed wind farm (VSWF 1) at bus #9, the variable speed wind farm VSWF2 shall be located at bus #32, the fixed speed wind farm (and its associated SVC) shall be located at bus #22 and the two additional SVCs are to be located at buses number #34 and #28 respectively with a var rating of 5Mvar each one as deduced from Table 6.8.

Figure 6.9 shows the network voltage profile under different load power demand and under wind power production. This configuration is the only one that keeps voltage profile within $\pm\, 5\% U_N$, for every sceneries. The other ones do not allow maintaining the voltage profile according to the standard voltage regulation.

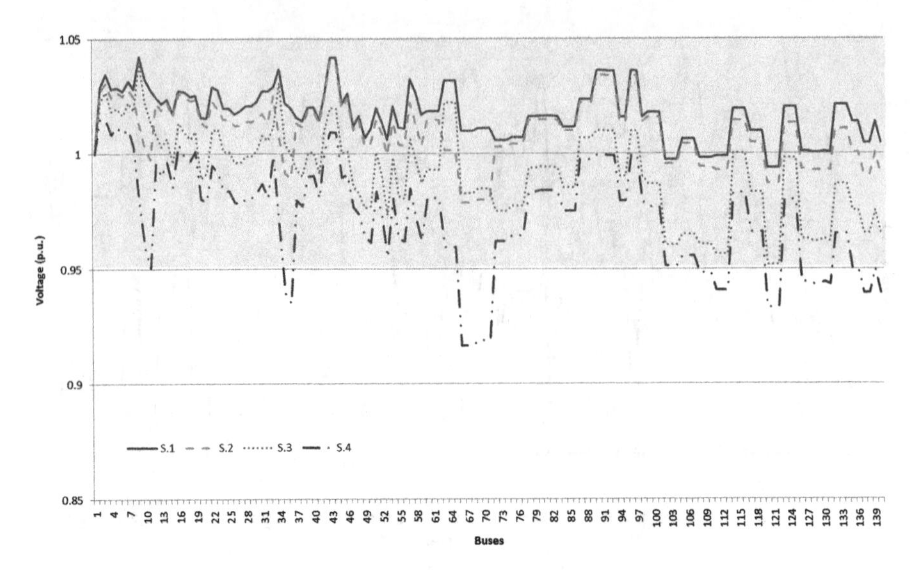

Fig. 6.8 Voltage profile considering the optimal configuration for scenario 3

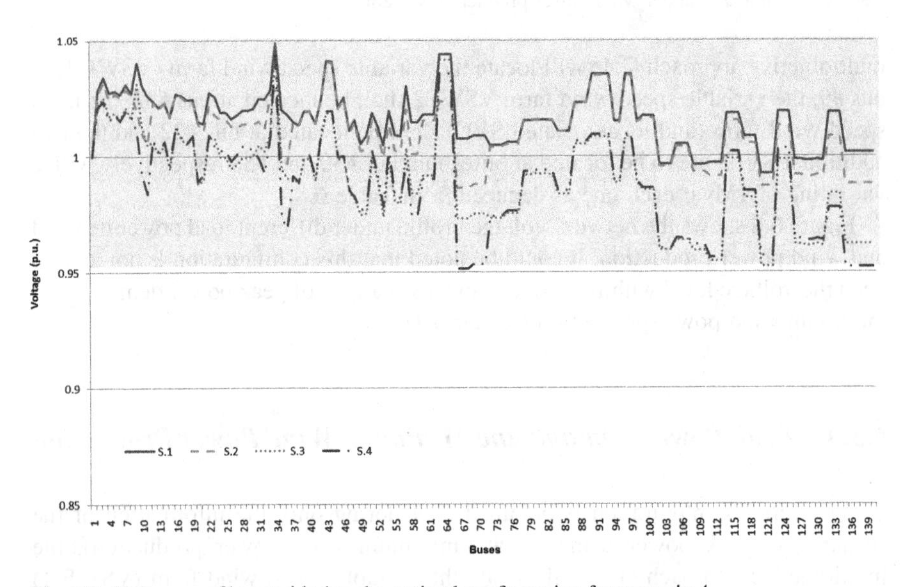

Fig. 6.9 Voltage profile considering the optimal configuration for scenario 4

6.4 Reactive Power Dispatch

In this case, the starting point relies on the network configuration given by the scenario of peak load and minimum generation shown in Table 6.9. Results of load-generation scenario number 4, which corresponds to the optimum network

Table 6.9 Results of load-generation scenario number 4

	PC1	PC2	PC3	Q3 (Mvar)	PC4	Q4 (Mvar)	PC5	Q5 (Mvar)
S.4	9	32	22	5	34	3	28	5

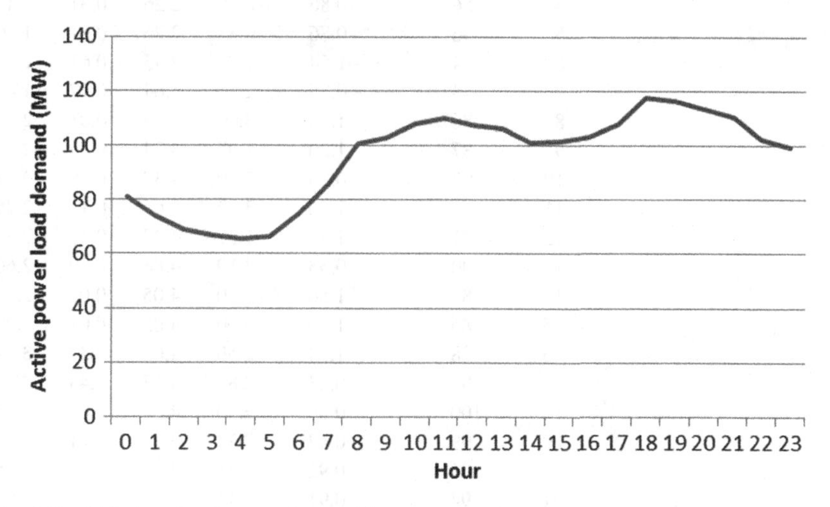

Fig. 6.10 24-h day-ahead power demand forecast

configuration in terms of all load scenarios (section 3.4). In this case, there are two variable-speed wind farms of 10 MW, each having a power factor of ± 0.95 p.u. and connected to nodes #9 and #32, and there is a 10 MW wind farm with fixed-speed machines and an associated SVC of 5 Mvar, which are connected to node #22. Moreover, the system relies on two SVCs of 3 and 5 Mvar connected to nodes #34 and #28, respectively.

For this study, for instance, a 24 h day-ahead forecast of the power demand provided by the Transmission System Operator is considered as is shown in Fig. 6.10.

It is important to note that at this point, the objective of the GA´s optimisation function is neither to determine the optimum location of the wind parks and SVCs, nor to set the maximum capacity thereof. Here, the optimisation method will be employed to operate the reactive power set points of the already installed wind parks and SVCs for the next 24 h. Table 6.10 shows the results of the reactive power management. First column represents the hourly period, the second column represents the percentage of load demand, columns three to seven show reactive power injection from VSWF wind farms (columns three and four), from SVC associated to the fixed speed wind farm (column five), and from the independent SVCs devices (columns six and seven).

A brief review of table shows that:

- For lowest demand, VSWF wind farms absorb reactive power. In the case of VSWF1, connected at bus #9, reactive power absorption is needed in the period from hour 0 to 6, and for VSWF2, allocated in bus #32, the absorption of reactive power is found to be between hour 0 and 3.

Table 6.10 Reactive power set points for the curve of the daily load

Hour	Load (%)	Q_1	Q_2	Q_3	Q_4	Q_5
0	69	−1.73	−1.43	3.18	1.04	3.95
1	63	−0.72	2.61	1.90	0.25	0.39
2	59	−2.62	1.88	2.36	0.93	2.48
3	57	−1.43	−0.17	2.73	0.63	1.26
4	56	−0.86	0.38	2.26	0.31	1.40
5	56	−0.86	0.38	2.26	0.31	1.40
6	64	−1.54	1.73	2.95	0.61	1.34
7	73	−0.85	2.23	3.64	0.77	1.60
8	86	1.50	3.00	4.05	0.01	2.30
9	87	1.20	1.97	3.74	0.36	2.04
10	92	0.83	2.80	4.37	0.48	2.28
11	94	1.20	3.12	4.67	0.17	2.75
12	91	1.31	1.37	4.27	0.17	2.57
13	90	0.53	3.19	4.88	0.79	2.60
14	86	1.50	3.00	4.05	0.01	2.30
15	86	1.50	3.00	4.05	0.01	2.30
16	88	1.24	2.50	4.03	0.34	3.04
17	92	0.83	2.80	4.37	0.48	2.28
18	100	0.80	2.00	4.62	0.67	3.77
19	99	0.47	2.77	4.72	1.21	3.10
20	96	0.92	2.41	4.81	0.64	2.65
21	94	0.91	2.11	4.44	0.80	3.07
22	87	1.20	1.97	3.74	0.36	2.04
23	85	−1.29	1.82	4.42	1.37	1.83

- In all 24-hourly periods the maximum reactive power delivered by wind farms at SVC devices, is inferior to their maximum capacity. So it is possible to schedule dynamically the reactive power injections from these devices according to the reactive power demand variations during the 24 h.
- At the same time, it can be deduced that the wind farms and SVC's optimally located and sized in Table 6.9, have the ability to offer a reactive power reserve in order to cope with voltage or abnormal load demand situations.
- It can be seen that the maximum reactive power injection for the different VAR sources is reached at different hourly periods. For example, VSWF1 injects its maximum reactive power value at hour 8, 14 and 15 (which corresponds to 1.5 MVAr); the maximum reactive power injection of VSWF2 and SVC associated to fixed speed wind farm is at hour 13 (3.19 MVAr and 4.88 MVAr, respectively). Finally SVC1 and SVC2 units inject their maximum reactive power at hour 23 and 0, respectively.
- Maximum active power load demand corresponds to hour 18. It must be noted that, for this condition, none of the wind farms or SVC units is injecting their maximum reactive power injection capability. So, optimization strategy is capable to schedule, in the worst loading situation, reactive power injection of different VAR sources without stressing any of them.

References

1. Raoufi H, Kalantar M (2009) Reactive power rescheduling with generator ranking for voltage stability improvement. Energ Convers Manage 50(4):1129–1135
2. Hugang X, Haozhong C, Haiyu L (2008) Optimal reactive power flow incorporating staticvoltage stability based on multi-objective adaptive immune algorithm. Energ Convers Manage 49(5):1175–1181
3. Miller TJE (1982) Reactive power control in electric systems. John Wiley & Sons, New York
4. Zhang W, Li F, Tolbert LM (2007) Review of reactive power planning: objectives, constraints, and algorithms. IEEE Trans Power Syst 22:2177–2186
5. Gitizadeh M, Kalantar M (2009) A novel approach for optimum allocation of facts devices usingmulti-objective function. Energ Convers Manage 50(3):682–690
6. Lahacani NA, Aouzellag D, Mendil B (2010) Contribution to the improvement of voltageprofile in electrical network with wind generator using svc device. Renew Energ 35(1):243–248
7. Wilch M, Pappala VS, Singh SN, Erlich I (2007) Reactive power generation by dfig basedwind farms with ac grid connection. In: Power Tech 2007 I.E. Lausanne, Switzerland, pp 626–632
8. Moghaddas-Tafreshi SM, Mashhour E (2009) Distributed generation modeling for powerflow studies and a three-phase unbalanced power flow solution for radial distribution systems considering distributed generation. Electric Power Syst Res 79(4):680–686
9. Braun M (2008) Reactive power supply by distributed generators. In: Proceedings of the IEEE power energy society general meeting, Calgary, pp 1–8
10. Sangsarawut P, Oonsivilai A, Kulworawanichpong T (2010) Optimal reactive power planning of doubly fed induction generators using genetic algorithms. In: Proceedings of the 5thIASME/WSEAS international conference on energy, Cambridge, UK, pp 278–282
11. Zhao M, Chen Z, Blaabjerg F (2009) Load flow analysis for variable speed offshore wind farms. IET Renew Power Gener 3(2):120–132
12. Hedayati H, Nabaviniaki SA, Akbarimajd A (2008) A method for placement of DG units in distribution networks. IEEE Trans Power Deliv 23(3):1620–1628
13. Lee KY, Bai X, Park Y-M (1995) Optimization method for reactive powerplanning by using a modified simple genetic algorithm. IEEE Trans Power Syst 10(4):1843–1850
14. Ajjarapu V (2006) Computational tecnhiques for voltage stability assessment and control. Springer, United States of America
15. Dai C, Chen W, Zhu Y, Zhang X (2009) Seeker optimization algorithm for optimal reactive power dispatch. Power Syst IEEE Trans 24(3):1218–1231
16. Pudjianto D, Ahmed S, Strbac G (2002) Allocation of var support using lp and nlp based optimal power flows. IEE Proc Gener Trans Distrib 149(4):377–383

Appendix

This appendix contains the electrical data of the 140 bus power network used in Chap. 6.

Load Data at Buses

Bus	Pd (p.u.)	Qd (p.u.)
1	0	0
2	0	0
3	24	−16
4	0	0
5	0	0
6	0	0
7	0	0
8	0	0
9	0	0
10	0	0
11	0	0
12	0	0
13	0	0
14	0	0
15	0	0
16	0	0
17	0	0
18	0	0
19	0	0
20	2.7	0.3
21	6.3	3.052
22	3.106	1.323
23	4.5	0.5

(continued)

H. Amaris et al., *Reactive Power Management of Power Networks with Wind Generation*, Lecture Notes in Energy 5, DOI 10.1007/978-1-4471-4667-4,

Bus	Pd (p.u.)	Qd (p.u.)
24	0	0
25	0	0
26	0	0
27	0	0
28	0	0
29	0	0
30	0	0
31	0	0
32	9	1
33	5.068	1.5237
34	0.2667	0.1997
35	0.3945	0.2147
36	0.5518	0.1472
37	0.627	0.1675
38	0.3793	0.1128
39	0.2521	0.0653
40	0.2521	0.0653

Bus	Pd (p.u.)	Qd (p.u.)
41	0.698	0.1776
42	0.2382	0.0658
43	0.428	0.1
44	0.428	0.108
45	0.428	0.108
46	1.0177	0.5493
47	0.8785	0.322
48	1.573	0.5169
49	0.626	0.2477
50	0.3723	0.191
51	0.0027	0
52	0.7479	0.1973
53	0.8183	0.373
54	0.0037	0.0023
55	0.3447	0.167
56	0.3447	0.167
57	3.0113	0.6113
58	1.4473	0.4757
59	1.4473	0.4757
60	3.3789	1.0158
61	3.3789	1.0158
62	3.3789	1.0158
63	0.1778	0.1332
64	0.1778	0.1332
65	0.1778	0.1332
66	0.2631	0.1351
67	0.2631	0.1351
68	0.2631	0.1351

(continued)

Bus	Pd (p.u.)	Qd (p.u.)
69	0.3678	0.1981
70	0.3678	0.1981
71	0.3678	0.1981
72	0.4175	0.226
73	0.4175	0.226
74	0.4175	0.226
75	0.2528	0.1962
76	0.2528	0.1962
77	0.2528	0.1962
78	0.168	0.0812
80	0.168	0.0812

Bus	Pd (p.u.)	Qd (p.u.)
81	0.168	0.0812
82	0.168	0.0812
83	0.168	0.0812
84	0.4651	0.1941
85	0.4651	0.1941
86	0.4651	0.1941
87	0.1588	0.0981
88	0.1588	0.0981
89	0.1588	0.0981
90	0.2854	0.1124
91	0.2854	0.1124
92	0.2854	0.1124
93	0.2854	0.1124
94	0.2854	0.1124
95	0.2854	0.1124
96	0.2854	0.1124
97	0.2854	0.1124
98	0.2854	0.1124
99	0.6784	0.3662
100	0.6784	0.3662
101	0.6784	0.3662
102	0.5916	0.322
103	0.5916	0.322
104	0.5916	0.322
105	1.0407	0.3447
106	1.0407	0.3447
107	1.0407	0.3447
108	0.4178	0.1651
109	0.4178	0.1651
110	0.4178	0.1651
111	0.2483	0.1273
112	0.2483	0.1273
113	0.2483	0.1273
114	0.0018	0.0014

(continued)

Bus	Pd (p.u.)	Qd (p.u.)
115	0.0018	0.0014
116	0.0018	0.0014
117	0.4985	0.2551
118	0.4985	0.2551
120	0.5456	0.2487

Bus	Pd (p.u.)	Qd (p.u.)
121	0.5456	0.2487
122	0.5456	0.2487
123	0.0024	0.0016
124	0.0024	0.0016
125	0.0024	0.0016
126	0.2298	0.1114
127	0.2298	0.1114
128	0.2298	0.1114
129	0.2298	0.1114
130	0.2298	0.1114
131	0.2298	0.1114
132	2.0076	0.4076
133	2.0076	0.4076
134	2.0076	0.4076
135	0.9649	0.3171
136	0.9649	0.3171
137	0.9649	0.3171
138	0.9649	0.3171
139	0.9649	0.3171
140	0.9649	0.3171

Line Data

From bus	To bus	R (p.u.)	X(p.u.)	B(p.u.)
3	7	0.0359	0.08015	0.01548
7	2	0.0274	0.0927	0.0197
2	4	0.01165	0.0394	0.00837
2	5	0.0329	0.11129	0.0236
2	6	0.0219	0.0742	0.0157
8	9	0.3468	0.578	0.00166
9	10	0.37	0.5916	0.00186
10	11	0.287	0.458	0.00144
8	12	0.1487	0.3887	0.0013
12	13	0.2574	0.331	0.00099
12	14	0.3634	0.4668	0.00141
14	15	0.375	0.3619	0.001

(continued)

From bus	To bus	R (p.u.)	X(p.u.)	B(p.u.)
14	16	0.122	0.2506	0.00077
16	17	0.122	0.2506	0.00077
17	12	0.1487	0.3887	0.0013

From bus	To bus	R (p.u.)	X (p.u.)	B (p.u.)
17	18	0.1416	0.3702	0.00123
18	19	0.122	0.237	0.00082
19	20	0.2533	0.5205	0.0016
19	20	0.909	0.6035	0.0017
19	21	0.23	0.778	0.00223
21	22	0.2332	0.2623	0.000856
22	23	0.2439	0.4888	0.00158
8	24	0.3831	0.4309	0.0014
24	25	0.003	0.00399	0.0000115
24	26	0.1363	0.1764	0.000525
24	27	0.3331	0.3747	0.0012
27	28	0.1261	0.1182	0.00035
28	29	0.00187	0.00385	0.0000118
28	30	0.10506	0.0985	0.00029
30	23	0.1998	0.225	0.00073
23	32	0.1514	0.1945	0.00059
32	31	0.042	0.0394	0.00116
23	31	0.0375	0.0794	0.00023
23	31	0.0375	0.0794	0.00023
1	2	0.00145	0.07345	0
4	23	0.00683	0.20483	0
5	16	0.01467	0.4163	0
47	19	0.106	1.4762	0
7	8	0.01467	0.34	0
7	8	0.01467	0.34	0
6	19	0.0073	0.1748	0

From bus	To bus	R (p.u.)	X (p.u.)	B (p.u.)
47	19	0.106	1.4762	0
7	8	0.01467	0.34	0
7	8	0.01467	0.34	0
7	33	0.01467	0.3263	0
34	9	0.106	1.476	0
10	35	0.232	2.388	0
36	11	0.106	1.596	0
12	37	0.106	1.496	0
38	13	0.106	1.596	0
14	39	0.106	1.756	0
14	40	0.106	1.596	0
15	41	0.0267	0.5327	0
16	42	0.044	0.798	0
17	43	0.106	1.596	0

(continued)

From bus	To bus	R (p.u.)	X (p.u.)	B (p.u.)
17	44	0.106	1.596	0
17	45	0.106	1.596	0
46	18	0.044	0.798	0
20	48	0.044	0.758	0
49	21	0.106	1.596	0
50	22	0.232	4.712	0
25	51	0.536	5.896	0
26	52	0.0267	0.5327	0
53	27	0.106	1.596	0
29	54	0.536	4.768	0
55	30	0.232	3.1508	0
56	30	0.232	3.312	0
32	58	0.06	0.9573	0
32	59	0.06	0.9573	0
31	57	0.0267	0.5193	0
33	60	0.3472	0.6014	0
33	61	0.3472	0.6014	0
33	62	0.3472	0.6014	0
34	63	2.0833	3.6083	0
34	64	2.0833	3.6083	0
34	65	2.0833	3.6083	0
35	66	4.167	7.2167	0
35	67	4.167	7.2167	0
35	68	4.167	7.2167	0
36	69	2.0833	3.6083	0

From bus	To bus	R (p.u.)	X (p.u.)	B (p.u.)
36	71	2.0833	3.6083	0
37	72	2.0833	3.6083	0
37	73	2.0833	3.6083	0
37	74	2.0833	3.6083	0
38	75	2.0833	3.6083	0
38	76	2.0833	3.6083	0
38	77	2.0833	3.6083	0
39	78	2.0833	3.6083	0
39	79	2.0833	3.6083	0
39	83	2.0833	3.6083	0
40	80	2.0833	3.6083	0
40	81	2.0833	3.6083	0
40	82	2.0833	3.6083	0
41	84	0.6944	1.2028	0
41	85	0.6944	1.2028	0
41	86	0.6944	1.2028	0
42	87	1.04167	1.8042	0
42	88	1.04167	1.8042	0
42	89	1.04167	1.8042	0
43	90	2.0833	3.6083	0

(continued)

From bus	To bus	R (p.u.)	X (p.u.)	B (p.u.)
43	91	2.0833	3.6083	0
44	92	2.0833	3.6083	0
44	93	2.0833	3.6083	0
45	94	2.0833	3.6083	0
45	95	2.0833	3.6083	0
43	96	2.0833	3.6083	0
44	97	2.0833	3.6083	0
45	98	2.0833	3.6083	0
46	99	1.04167	1.8042	0
46	100	1.04167	1.8042	0
46	101	1.04167	1.8042	0
47	102	2.0833	3.6083	0
47	103	2.0833	3.6083	0
47	104	2.0833	3.6083	0
48	105	1.04167	1.8042	0
48	106	1.04167	1.8042	0
48	107	1.04167	1.8042	0
49	108	2.0833	3.6083	0
49	109	2.0833	3.6083	0
49	110	2.0833	3.6083	0

From bus	To bus	R (p.u.)	X (p.u.)	B (p.u.)
50	111	4.167	7.2167	0
50	112	4.167	7.2167	0
50	113	4.167	7.2167	0
51	114	8.33	14.43	0
51	115	8.33	14.43	0
51	116	8.33	14.43	0
52	117	0.6944	1.2028	0
52	118	0.6944	1.2028	0
52	119	0.6944	1.2028	0
53	120	2.0833	3.6083	0
53	121	2.0833	3.6083	0
53	122	2.0833	3.6083	0
54	123	8.33	14.43	0
54	124	8.33	14.43	0
54	125	8.33	14.43	0
55	126	4.167	7.2167	0
55	127	4.167	7.2167	0
56	128	4.167	7.2167	0
56	129	4.167	7.2167	0
55	130	4.167	7.2167	0
56	131	4.167	7.2167	0
57	132	0.6944	1.2028	0
57	133	0.6944	1.2028	0
57	134	0.6944	1.2028	0
58	135	1.389	2.405	0

(continued)

From bus	To bus	R (p.u.)	X (p.u.)	B (p.u.)
58	136	1.389	2.405	0
59	137	1.389	2.405	0
59	138	1.389	2.405	0
58	139	1.389	2.405	0
59	140	1.389	2.405	0